Birds of the
Lahontan Valley

Loggerhead Shrike

Birds of the Lahontan Valley

A GUIDE TO NEVADA'S WETLAND OASIS

Graham Chisholm
and Larry A. Neel

ILLUSTRATIONS BY
MIMI HOPPE WOLF

University of Nevada Press
Reno & Las Vegas

This book was funded in part by grants from
the Great Basin Bird Observatory,
the William H. and Mattie Wattis Harris Foundation,
and the John Ben Snow Memorial Trust.

University of Nevada Press, Reno, Nevada 89557 USA
Copyright © 2002 by University of Nevada Press
Line art illustrations copyright © 2002 by Mimi Hoppe Wolf
Manufactured in the United States of America
Design by Kaelin Chappell
Library of Congress Cataloging-in-Publication Data
Chisholm, Graham, 1959–
Birds of the Lahontan Valley / Graham Chisholm and Larry A. Neel.
p. cm.
Includes bibliographical references (p.   ).
ISBN 0-87417-479-1 (pbk. : alk. paper)
1. Birds—Lahontan Valley.
2. Wetland ecology—Lahontan Valley.
I. Neel, Larry A. II. Title.
QL683.L18 C55 2001
598'.09793'52—dc21
2001002237
The paper used in this book meets the requirements of
American National Standard for Information Sciences—
Permanence of Paper for Printed Library Materials,
ANSI Z39.48-1984. Binding materials were selected
for strength and durability.
First Printing
11 10 09 08 07 06 05 04 03 02   5 4 3 2 1

To my parents,
Bill and Dorothy Chisholm

G.C.

To my wife, Martha,
who married me in a basketball gym
and has never once tired of my dreaming

L.N.

Contents

2. SPECIES ACCOUNTS

Illustrations

MAPS

Acknowledgments

This project grew out of an ongoing conversation about the Lahontan Valley's bird life and would not have been possible without the pioneering work of Ray Alcorn or the growing interest in restoring the Stillwater and Carson Lake wetlands. The latter is thanks in large part to a committed group of individuals, including Ron Anglin, Norm Saake, and Steve Thompson.

Since we started this project in December 1994, a number of people who deserve special mention have made concrete suggestions or contributed information. We take this opportunity to express our deep appreciation to the following friends and fellow birders: Ray Alcorn, Bill and Beth Clark, Gary Cottle, Chris Elphick, Robert Flores, William Henry, Anne Janik, David Jickling, Hugh Judd, Paul Lehman, Bill Mewalt, Martin Meyers, Dave McNinch, Margaret Rubega, Norm Saake, Greg Scyphers, Dennis Serdehely, Jane Thompson, Steve Thompson, Dennis Trousdale, Jack Walters, and Diane Wong.

A number of individuals deserve special mention for helping us understand the complex biological, hydrologic, social, and economic issues surrounding the Lahontan Valley wetlands and the Newlands Project. Their diverse views and perspectives have generated a vigorous forum in which we have had the privilege of participating and from which we have learned much. At times those views and perspectives have clashed, but through it all we have developed the deepest respect for many of the principal spokespersons. For the depth of their convictions and willingness to share their knowledge we thank Ted DeBraga, Cliff Creger, Jim Curran, Carl Dodge, John Doebel, Connie Douglas, Kay Fowler, Dennis Ghighlieri, Jim Giudici, Rich Heap, Jim Johnson, Dave Livermore, Lyman McConnell, Roger Mills, William Molini, Tina Nappe, Mary Reid, Betsy Rieke, Gary Shellhorn, William Sheppard, Rose Strickland, Don Travis, Robert Wigington, and David Yardas.

A number of museums allowed us to examine their collections, and we are grateful for their assistance: Don Baepler, Barrick Museum of Natural History, Las Vegas, Nevada; Alan Gubanich, Museum of Vertebrate Zoology, Reno, Nevada; Ned K. Johnson and Carla Cicero, Museum of Vertebrate Zoology, Berkeley, California; Jim Dean, U.S. National Museum of Natural History; and Mark Robbins, Natural History Museum, Lawrence, Kansas.

We owe a big debt to several people who improved the text considerably with their editorial patience and expertise: Don DeLong, David Holway, Dave Marshall, and Peter Wigand.

Finally, we must acknowledge the debt we owe to our families—Kelly Cash and Ian and Lee Chisholm, and Martha Neel and Darcy Ward—without whom all this would be for nothing.

Introduction

The Lahontan Valley wetlands of the Carson River are among the most important, and most threatened, wetland systems in the intermountain West and have rightly received international recognition as a critical steppingstone on the Pacific Flyway. These terminal wetlands, which developed approximately 4000 years ago, have experienced dramatic changes in the past 100 years. Agricultural water diversions, the creation of Lahontan Reservoir, and the development of the Comstock Lode mining district substantially reduced the size and quality of the wetlands. A recent estimate placed the acreage of the Lahontan Valley wetlands, including the Carson Sink, at an average of 150,000 acres in the period 1845–60. Today, wetland acreage is estimated to average less than 10,000 acres.

The purpose of this book is to bring together the wealth of information that exists on the bird life of the Lahontan Valley. We were fortunate to be able to draw on the observations of amateur naturalists and scientists who over the past century have taken careful notes on bird species and habitat in the valley. We begin with a discussion of the valley's natural and human history, particularly the story of the Lahontan Valley wetlands. The species accounts that follow in chapter 2 build on Ray Alcorn's works, especially *Birds of Nevada,* to provide the first comprehensive review of the Lahontan Valley's avian life. This book provides detailed status, habitat, and seasonal occurrence information for the 297 bird species recorded in the valley. The appendixes include nesting records for colony-nesting wading birds and seasonal shorebird counts (appendix 1), and Christmas bird counts for the period 1985–99 (appendix 2). The book also includes a site guide to the valley's most significant birding sites.

The recent efforts of the U.S. Fish and Wildlife Service, the state of Nevada, and private groups to restore the wetlands through the acquisition of water rights are breathing fresh life into Stillwater Marsh and the Carson Lake wetlands. The growing interest in riparian habitat may help restore the lower Carson River corridor. Although far from complete, this review can serve as a baseline from which to monitor future trends in avian populations and the changes occurring in the wetlands and elsewhere in the valley. We hope this book will help spark greater interest in both the Lahontan Valley's bird life and the habitats on which the birds depend.

I.
The Lahontan Valley

GEOGRAPHIC SCOPE

The Lahontan Valley, also known as the Carson Desert, covers 1383 square miles in northern Nevada. The valley's northern boundary is formed by the West Humboldt Mountains, a small range that separates the Carson Sink from the Humboldt Sink to the north. To the east, the Stillwater Range rises to 8790 feet—nearly 5000 feet above the valley's low point of 3860 feet above sea level. The valley's southern boundary is marked by a series of discontinuous low mountains, including the Cocoon Mountains, the White Throne Mountains, and the Desert Mountains. The western boundary is somewhat indefinite, lying just to the east of the Virginia Range.

The geographic coverage of this book is limited to the portions of the Lahontan Valley that lie east of a line extending from Lahontan Reservoir through the town of Hazen. Our coverage focuses on the valley's wetlands—the most prominent natural feature—and their associated upland communities; we do not cover the surrounding ranges. For the most part, we have confined our scope to the valley below Lahontan Dam, although in certain instances where it is relevant, sightings from the lower end of Lahontan Reservoir are included.

CLIMATE

The Lahontan Valley is one of the warmest and driest basins in northern Nevada because of its low altitude (4000 feet average) and its location in the Sierra Nevada's rain shadow. The mean annual

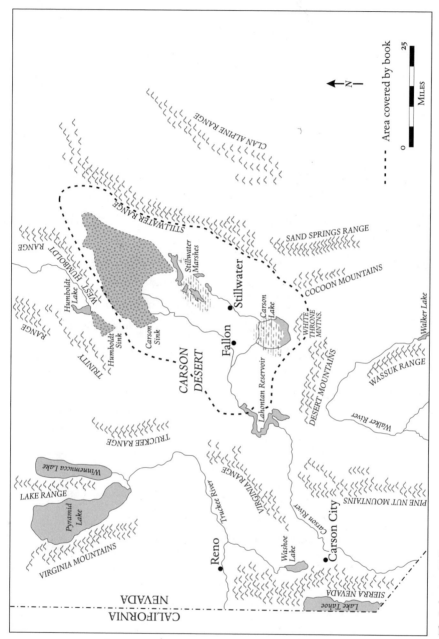

The Lahontan Valley

precipitation recorded at Fallon is 5.32 inches. The valley's precipitation is primarily linked to two types of storms: Pacific lows and Great Basin, or Tonopah, lows. Pacific storms typically bring most of the winter precipitation, whereas the Great Basin lows generate storms that bring precipitation in the spring (Houghton et al. 1975). It is not unusual for the valley to experience precipitation extremes between 1.6 and 9 inches annually.

The maximum temperature recorded in Fallon is 107°F, and the minimum is –27°F. The maximum average temperatures are 91°F in July and 44°F in January. Fifty-degree temperature variations are not uncommon during summer days. As a desert, the valley is typified by low humidity and high evaporation rates (approximately 60 inches annually; J. Ashby, Western Regional Climate Center, Desert Research Institute, pers. comm. 1996).

GEOLOGIC HISTORY

The Lahontan Valley was inundated during the Pleistocene by Lake Lahontan, an 8665-square-mile body of water that covered much of western and northern Nevada. Lake Lahontan reached its high point of 4363 feet above sea level approximately 14,000 years ago. The lake's level has varied greatly over the past 50,000 years, and several distinct beach lines are visible on the mountains and hills that surround the valley. Approximately 10,000 years ago Lake Lahontan was becoming a shallow lake and the Lahontan Valley wetlands were emerging.

Sediment from Lake Lahontan—clays, gravels, and sands—covers much of the Lahontan Valley below the lake's highest level. In places, two additional deposits overlie the lake sediment: light-colored aeolian sands that form a system of dunes, most of which are covered with vegetation; and clays, silts, and salts deposited by the Carson River's inflow into the valley, including deposits left behind by evaporation (Morrison 1964).

Unlike much of the Great Basin, the Lahontan Valley is not typified by a sea of sagebrush. Large areas of the Carson Desert, particularly the drier sites associated with gravelly and well-drained sediments of Lake Lahontan, as well as the surrounding hills above the valley floor, are covered by shadscale (*Atriplex*); small, grayish Bailey's greasewood (*Sarcobatus baileyi*); and low, spiny bud sage (*Artemisia spinescens*). These low woody shrubs are widely spaced, typically leaf out in April, and bloom, set seed, and go dormant before the onset of hot weather in mid-June. Winterfat (*Ceratoides lanata*) can be found in this association as well (Billings 1945; Grayson 1993).

Stabilized sand dunes often support significant stands of smokebush, or Nevada dalea (*Psorothamnus polydenius*), a plant that had a variety of medicinal uses among the Northern Paiute. Four-winged saltbush (*Atriplex canescens*) is commonly found with smokebush on these dunes, as are two species of horsebrush (*Tetradymia*), Indian ricegrass (*Oryzopsis hymenoides*), and rabbitbrush (*Chrysothamnus nauseosus*). These dunes are typical of the country extending from Indian Lakes to Timber Lakes and eastward.

The low, poorly drained alkaline soils around playas are dominated by extensive stands of big greasewood (*Sarcobatus vermiculatus*) and large-leaved Torrey saltbush (*Atriplex torreyi*). These two dominant species are joined by pickleweed (*Allenrolfea occidentalis*), quail brush (*Atriplex lentiformis*), and other salt-tolerant species (Fowler 1992; Grayson 1993; Mozingo 1987).

The only big sagebrush (*Artemisia tridentata*) community in the Lahontan Valley occurs along the Carson River floodplain downstream from Lahontan Reservoir to the Carson River Diversion Dam. Sagebrush replaces the Bailey's greasewood–shadscale association as elevation increases in the Stillwater Range.

The Carson River corridor historically was dominated by cottonwood (*Populus fremontii*) and willows, but in recent years the understory has been invaded by Russian olive (*Elaeagnus angustifolia*) and tamarisk, or salt-cedar (*Tamarix*). These two aggressive exotics have affected many western rivers, encouraged by changes in flood regime associated with controlled water diversions and flow manipulations. Formerly, floods reduced soil salinity and helped recruit a new generation of cottonwoods, which are today commonly found around farmyards. Willows (*Salix* spp.) are still common on many drains and ditches. Kay Fowler's interviews with Wuzzie George, a Northern Paiute woman from Fallon, point to the decline of silver buffaloberry (*Shepherdia argentea*), another important riparian species that now occurs only in isolated patches in the Lahontan Valley and upstream of Lahontan Reservoir on the Carson River. At one time, buffaloberry was common along the lower Carson River corridor and occurred along Stillwater Slough (Fowler 1992).

The loss and degradation of riparian habitat along the lower Carson River has occurred over a long period and is linked to upstream impoundment, changing water flows, livestock grazing, clearing of riparian forest, channelization, development in the floodplain, and other practices associated with human habitation. The loss or degradation of the cottonwood-willow riparian forest in large portions of the valley probably contributed to the partial or complete loss of certain bird species there.

The Lahontan Valley wetlands include diverse wetland communities ranging from fresh to saline and from wet sheens to relatively deep-water habitat. The predominant wetland plants include alkali bulrush (*Scirpus maritimus*), hardstem bulrush (*Scirpus acutus*), two species of cattails (*Typus* spp.), sago pondweed (*Potamogeton pectinatus*), wigeon grass (*Ruppia maritima*), and arrowleaf (*Sagittaria cuneata*). A recent survey located 21 of the 27 wetland plant associations historically described for the Lahontan Valley despite the substantial transformation of the wetlands. Further, the survey found that the diagnostic species for another 4 associations remained present but no longer dominated the sampled sites (Bundy et al. 1996). The continued presence of so many of the historical wetland plants is cause for great optimism regarding the potential for wetland restoration.

THE END OF THE PLEISTOCENE

A person standing 13,000 years ago at the mouth of one of the caves in the north face of the White Thrones, which overlook the Lahontan Valley from the south end of Carson Lake, would have witnessed a very different scene from the one visible today. The entire valley was under water from the White Thrones to the long peninsula of the Humboldt Range to the north, and even beyond. It was a deep lake, much like Pyramid Lake is today, with a fairly rapidly falling littoral zone that was not conducive to forming marshes. Rattlesnake Hill was an island similar to Anaho Island in Pyramid Lake.

The sky above that ancient observer was very likely cloudy, and the temperature was probably a brisk 50°F. The icepack far to the north (although only a traveler of Odyssean caliber would have known about it) held the Arctic storm fronts at bay, creating a milder, if somewhat cooler, climate than the one we know today. The observer probably took note during his lifetime, however, that the summers were getting hotter and the winters were getting colder. The northern ice packs were receding, withdrawing their stabilizing influence on the weather as they went.

The hills above were covered with a mix of sagebrush and saltbush. At some places along the shore, juniper woodland extended down from the fans and foothills above. Not such a dramatic change, really; not so different from many of the ranges of central Nevada today. The ancients shared those uplands, however, with a very different assemblage of wildlife than the one that inhabits it today. The grassy benches, such as the tables of the Stillwater Range under Job's Peak, supported herds of endemic horses and camels. The species of pronghorn we know today was there, joined by a diminutive cousin only 2 feet tall at the shoulder that weighed only 20 pounds and sported a 4-pronged horn core rather than the 2-pronged core of our present species. It is possible that during their wanderings the valley's Pleistocene inhabitants encountered one of the last of the Columbian mammoths, a megabeast that was winding down its impressive existence on the planet. We know that mammoths came down to the northern beaches of Lake Lahontan (Pyramid Lake and the Black Rock Desert) for a drink, and sometimes got mired in the mucky margins of the lake. There is no hard evidence that these hairless relatives of the more famous wooly mammoth were also roaming the hills south of the giant lake, but the probability of that is high. There is also good evidence that a species of bison roamed these ranges as well.

Scientists strongly suspect that the herds of herbivores were enormous, because the diversity of carnivores and scavengers that depended on them for food was high. Saber-toothed cats stalked the valley, as did short-faced bears, fearsome predators larger and faster than grizzlies. Dire wolves and dirk-toothed cats (comparable to today's bobcat) lurked in the blackness beyond the campfires at night, in the hope that dinner scraps would be left behind.

The skies above were patrolled by several great scavenger birds. In addition to the turkey vultures and black vultures we know today (or forms very similar to them), there were California condors and a slightly larger version now known as Clark's condor. Bigger than both was Merriam's teratorn, with a wingspan of 12 feet and a long, hooked beak that probably made it look something like today's crested caracara. And if that wasn't enough, an even bigger teratorn with a wingspan greater than 17 feet ruled the skies, and probably had its way at the carcasses that betrayed the work of the great cats that followed the herds.

Marshes developed on the shallow fringes of the lake, most likely on the north end where it disappeared behind the Humboldt Range. The bird life was fairly similar to what we know today, with a few notable additions. Two species of flamingos, one larger than today's greater flamingo and one about three-quarters its size, frequented the mudflats on the north end. White pelicans, cormorants, and gulls nested on Rattlesnake Hill much as they do today on Anaho Island. During migration, loons and western grebes joined the pelicans and cormorants for about a month at a time, feeding on the myriad schools of tui chub and the sucker the native people would later call cui-ui. The availability of these food resources was critical to the migratory birds, for it allowed them to rebuild their fuel reserves to power them on to breeding grounds farther to the north.

Ducks were mostly of the diver variety—canvasbacks, redheads, scaup, mergansers, and an occasional scoter. They rimmed the perimeter of the lake and rafted on the north end, exploiting the freshwater mussel beds that dotted the narrow littoral zone. Perhaps they were joined by flocks of small pygmy geese no larger than mallards. Dabbling ducks and the myriad shorebirds we are accustomed to seeing today were mostly confined to the mudflats and marshes on the shallow north end of the lake in what is now known as Lovelock Valley.[1]

TEN THOUSAND YEARS AGO

Could our ancient observer have returned and stood at the same vantage point 3000 years later, the scene would have been different yet again. The deep water had receded 10,000 years ago, leaving behind a mosaic of pools and vast shallow emergent wetlands. During the next

6000 years the water levels became much more dynamic, rising and falling with the destabilized climate, as well as with the fickle flows of the rivers, which evidently changed terminal basins sometimes.

Gone were the mammoths, sabertooths, little pronghorns, flamingos, and teratorns. Instead, endless flocks of waterfowl wheeled in the sky over the marshes to the north. Rattlesnake Hill was still a nesting island for pelicans, and the more prevalent shallow waters greatly facilitated their feeding. The air sparkled with string after string of white pearls . . . pelicans leaving the island after being relieved by their mates at the nest.

The vast tule stands between Carson Lake and the Carson Sink teemed with colony-nesting wading birds—white-faced ibis, great egrets, snowy egrets, black-crowned night herons. Great blue herons nested in dense rookeries in the cottonwood trees along the river delta on the west side of the valley.

The marshes literally swarmed with ducks—mallards, pintails, gadwalls, cinnamon teal, and, during migration, shovelers and green-winged teal. Tens of thousands of snow geese and their smaller cousins, Ross' geese, would descend on the salt-grass meadows and nut grass (alkali bulrush) margins as soon as the ice began to open up in February. The white swarm zeroed in on the previous summer's new growth, uprooting it to get at the tasty, nutritious tubers underground. A night's feeding by the geese would leave a young nut grass stand totally thrashed; and in that way, natural openings developed in the bulrush stands that favored duck brooding; the marsh was kept in a healthy, dynamic condition; and nutrients were cycled on a hemispheric scale.

The water was considerably saltier than it had been 3000 years earlier because of in situ evaporation and concentration of the great Pleistocene lake. Now, the water spreading across the great expanse of the Carson Sink, its ultimate destination, was at its saltiest, and only inches deep at its northernmost margin in most years. Here hundreds of thousands of shorebirds loaded up on the copious invertebrate blooms in the water and the mud. On blustery days, the wind would literally roll the water back like a giant carpet, allowing Arctic migrants to replenish their fat-fired fuel tanks on the newly exposed mudflats—dowitchers, sandpipers, plovers, phalaropes. Marbled godwits gorged themselves on water beetles in belly-deep water on their way to their prairie nesting grounds in what are now Montana, Saskatchewan, and Alberta. Yellowlegs stalked the marshy deltas. Staying behind to nest on the sandy islands along the margins of the sink were American avocets and black-necked stilts by the thousand. Long-billed curlews and Wilson's phalaropes nested in the spikerush sloughs and salt-grass meadows of the Carson River delta. Way out to the north, just beyond the water's margin, balls of fluff would suddenly turn into snowy plovers and skitter across the blinding white expanse at the slightest threat. The magnitude of feathered biomass was overwhelming, to say nothing of the amount of invertebrate biomass converted to fat and recycled to the ends of the hemisphere. In that sprawling scene one could sense the breathing of the planet itself.[2]

WETLANDS

Since the last recession of Lake Lahontan, the Carson River has flowed from the Sierra Nevada to the south and east of Lake Tahoe, through the Carson Valley, to a large terminal delta in that portion of the Carson Desert known as the Lahontan Valley. This delta once comprised a vast wetland complex covering on average 150,000 acres, but probably contracting to as little as 25,000 acres and expanding to as much as 250,000 acres during the boom-and-bust cycles that characterize Sierra snowpacks (Kerley et al. 1993).

The Carson River generally flowed south in the Lahontan Valley into Carson Lake (at times referred to as South Carson Lake), though there is evidence that in the mid-1800s at least a portion of the Carson River also flowed northward into the Carson Sink (at times referred to as North Carson Lake; Kerley et al. 1993). It is likely that the Carson River channel naturally meandered and changed course over time.

When Europeans first reached the region, the wetlands formed where the Carson River flowed into South Carson Lake varied in size from 25,000 to 34,000 acres depending on drought cycles. Carson Lake was a relatively freshwater pluvial lake with a maximum depth of 10 feet. As spring runoff raised the lake level each year, water discharged into Stillwater Slough, a substantial channel on the lake's northeast side. In 1861, the slough was 60 feet wide, with 8- to 10-foot vertical banks (DeQuille 1963).

Stillwater Slough meandered some 12 miles to the northeast before discharging into a series of "lakes" called the Stillwater Marsh. The marsh was a patchwork of open water of variable depth combined with substantial stands of tules, and was slightly more saline than Carson Lake. The remains of freshwater clams, minks, and river otters in archaeological sites indicate that the marsh was once less saline than at present. Because the Carson River's flow through Stillwater Slough was seasonally intermittent, parts of Stillwater Marsh were subject to seasonal drying (Kerley et al. 1993).

During most winters, the Carson River's inflow would have filled Carson Lake and the Stillwater Marsh, causing the latter to overflow to the north into the Carson Sink. The Carson Sink in some years received a substantial flow from the Carson River and grew to cover more than 190,000 acres, as occurred in the 1860s and 1985–86. In most years, however, the Carson Sink wetlands were significantly smaller and subject to rapid drying as the summer evaporation rate increased. The ephemeral and shallow Carson Sink wetlands provided important playa, or mudflat, habitats. This playa system also played an important role in helping remove salts from the system.

Spring runoff would flush salts out of Carson Lake and Stillwater Marsh and deposit it out on the playa, where the water evaporated and winds carried away the remaining salts.

With some variations this natural system remained in place until a combination of natural and human events led to a dramatic transformation. Flooding in the winter of 1861–62 enlarged the north channel, and the flood-control measures taken in 1863 led to the creation of the New River channel during the 1867 flood (Anglin 1994).

THE CATTAIL-EATERS

For thousands of years the Cattail-Eaters, or Toidikadi Northern Paiute people, lived in a large area surrounding the wetlands of the Carson Desert. European settlement of the Carson Desert beginning shortly after 1850 led to the gradual displacement of the Cattail-Eaters from their traditional lands.

Before the Europeans arrived, the Cattail-Eaters relied on plant and animal life from the marshes for a majority of their needs. Although they foraged seasonally in the surrounding upland habitats, particularly in the Stillwater and Clan Alpine Ranges, the wetlands were the heart of their territory. The roots, leaves, seeds, and pollen of marsh vegetation—cattails, hardstem and alkali bulrushes, and other species—were key foods. In addition, marsh vegetation was used to create houses, boats, decoys, clothing, bags, baskets, mats, and other needed objects (Fowler 1992).

The Cattail-Eaters also relied on the marsh's water birds—including ducks, geese, swans, coots, pelicans, herons, and cormorants—for food. In addition to netting and hunting birds, they harvested eggs, which they stored in the cool, sandy mud. The marshes also provided tui chub, Tahoe sucker, redside shiners, and speckled dace, as well as freshwater mollusks and various insects. In the surrounding upland, the Cattail-Eaters caught ground squirrels, jackrabbits, cottontails, and other small game. Seasonal expeditions into the mountains yielded bighorn sheep and mule deer (Fowler 1992).

THE NEWLANDS PROJECT AND ITS
IMPACT ON THE WETLANDS

The advent of agriculture in the Lahontan Valley, initially based on the harvesting of natural meadow vegetation, gradually led to efforts to create ditches to deliver spring runoff from the river channels. Until the advent of the Newlands Project—the first federal reclamation project—beginning in 1903, the impacts of water diversion were relatively minor.

The Newlands Project, and particularly the completion of the Truckee Canal in 1905 and Lahontan Dam in 1915, doubled the valley's irrigated acreage and replaced natural haying practices with cultivation (Townley 1998). The project, especially the ability it gave the valley's settlers to divert and import large amounts of Truckee River water (an average of one-half of the Truckee River's flow) and impound large amounts of upstream flows, changed the timing, quality, and quantity of water flow into the wetlands. The waters below the dam were routed through a series of drains rather than flowing through natural channels out to the wetlands. Drain water from irrigated lands reached the wetlands, but drain flows were linked to the irrigation season and provided a more constant flow during the spring and summer months instead of the former higher spring flows. The Lahontan Valley wetlands also benefited from winter water releases from Lahontan Reservoir for hydropower generation (U.S. Fish and Wildlife Service [USFWS] 1996).

The Newlands Project and the changes in the wetlands hydrology it brought about altered the composition and extent of the wetland vegetation. For example, it is estimated that between 1900 and 1952, the acreage of hardstem bulrush was cut in half while the acreage of cattails almost tripled (Fowler 1992). The project's overall impact on the wetlands was minimized, however, by the importation of large amounts of Truckee River water. Between 1905 and 1967, when measures were initiated to reduce the amount of Truckee River water being diverted for the use of Newlands Project irrigators, the wetlands continued to provide exceptionally good habitat for breeding and migratory birds and supported a large trapping industry and fishery.

The Greenhead Hunting Club established a presence at Carson Lake very early in the history of the Newlands Project, probably as early as 1915, and entered into a formal agreement with the Truckee–Carson Irrigation District in 1928 for wetland development and waterfowl hunting. The demand for a high-quality duck-hunting area was instrumental in preserving wetlands at Carson Lake after the lake's designation as a reclamation project sump. The Greenhead Hunting Club became an active voice in the management of Carson Lake, along with the Truckee-Carson Irrigation District, and helped emphasize the importance of the wetlands to migratory waterfowl (USFWS 1996).

In 1931, the Fallon National Wildlife Refuge was established at the mouth of the Carson River to provide a refuge and breeding ground for birds and other wildlife. In 1948, the U.S. Fish and Wildlife Service, the Nevada Fish and Game Commission, and the Truckee-Carson Irrigation District entered into an agreement to jointly manage the Stillwater wetlands as a public shooting ground (State of Nevada 1950). This cooperative effort was galvanized by rumors that affluent duck hunters from out of state were mounting a campaign to convert the Stillwater marshes to private ownership. As the relatively new applied science of wildlife management shed its training wheels and burgeoned during the post–World War II economic boom, the primary management emphasis of the newly created Stillwater Wildlife Management Area became the construction of water management infrastructure. More than 30 miles of dikes, 70 miles of canals, and 200 water control structures were installed (State of Nevada 1953). Most of the wetland units south of Division Road in the refuge were created during this period. These changes completed a protracted progression from "natural" to "managed" wetlands.

The Newlands Project's drain water and seepage also created or enhanced wetlands such as Massie and Mahala Sloughs, Fernley Sink, and the Indian Lakes area. These new wetlands provide habitat of varying qualities, as do the project's regulatory reservoirs, including S-Line, Harmon, Sheckler, Old River, and Ole's Pond. On the other hand, the Newlands Project's diversion of large amounts of Truckee River water greatly contributed to the serious decline of Pyramid Lake and the eventual disappearance of Winnemucca Lake, once renowned as a migratory bird area.

In fact, it was the Newlands Project's impacts on Pyramid Lake that in 1967 led to the implementation of a series of federal regulations and court decrees known today as the Operating Criteria and Procedures (OCAP). OCAP began reducing Truckee River diversions by the Newlands Project in order to protect Pyramid Lake and its endangered cui-ui. OCAP specifically focused on eliminating winter hydropower flows and reducing farm runoff, spills, and conveyance losses. As these sources of water for the wetlands were reduced or lost, the wetlands' very existence became tenuous. The importation of Truckee River water had masked the impact of the Newlands Project and irrigation diversion along the upper Carson River on the Lahontan Valley wetlands. What had begun as remedial action to repair one environmental crisis was now inadvertently revealing another one just as tragic (Chisholm 1994; USFWS 1996; Yardas 1994).

Recognizing that OCAP had substantially decreased wetland acreage, wetland advocates searched for ways to build public support for the Lahontan Valley wetlands. This effort received a substantial boost with the 1988 designation of the Lahtonan Valley wetlands as a Western Hemispheric Shorebird Reserve Network site (Neel and Henry 1997). In 1988, the Lahontan Wetlands Coalition was formed, bringing together waterfowl hunters, environmentalists, and others to speak out on behalf of the wetlands. The coalition was instrumental in ensuring that Public Law 101-618 included a provision authorizing the restoration of a long-term average of 25,000 acres of wetlands through a water-rights acquisition program. Although the acquisition program started out slowly, the U.S. Fish and Wildlife Service, the state of Nevada, the Nevada Waterfowl Association, and The Nature Conservancy had purchased approximately 34,000 acre-feet of water rights by early 2001. The U.S. Fish and Wildlife Service has established a goal of purchasing at least 75,000 acre-feet of water rights. In addition, agricultural drain flow, spills from Lahontan Reservoir, purchases of upper Carson River water rights, and other measures are anticipated to provide an additional 50,000 acre-feet of water for the wetlands (USFWS 1996). The purchase of water rights allows the USFWS and the Nevada Division of Wildlife to order water just like any other irrigator. Dependable deliveries give wetland managers the ability to achieve habitat restoration goals in the most efficient manner.

WETLAND CONTAMINATION FROM MINING AND AGRICULTURE

During the Comstock mining era of the 1860s and later, thousands of tons of mercury waste were dumped into the Carson River from the gold-processing mills that dotted the floodplain from Mexican Dam to Dayton. This led to the contamination of river sediment all the way to the terminal wetlands of the Lahontan Valley (Tuttle and Thodal 1998). In addition, the desert soils of the valley are naturally high in other heavy metals—selenium, boron, molybdenum, and manganese, to name a few. Historically, these metals leached out of the soils as salts during floods, and these solutes found their way down to the terminal wetlands as well (Lico 1992; Tuttle and Thodal 1998). Flood irrigation magnified that process for some years after the initiation of the Newlands Project; indeed, farmers found it necessary to leach the soils of the bulk of their salts and keep them drained in order to maintain productive soil. The serious threat to ecological health posed by this process was brought to national attention in the 1980s when new agricultural drains were created to serve certain lands in the northern portion of the Newlands Project (Lico 1992). These new drains contributed elevated salt loads to the Stillwater marshes that threatened to seriously affect bird reproduction. Luckily, provisions of Public Law 101-618 called for the closure and filling of these new drains. Those provisions were quickly implemented, averting further damage. The concentration of heavy metal salts in the sediments of the primary wetlands of the Lahontan Valley continues to be of concern to biologists and conservationists. The only long-term treatment of that chronic threat involves the continual flushing of drain flows and sediments with prime water deliveries. Periodic draining and drying of lake bottoms also facilitate the removal of salt residues by the wind.

GREAT BASIN WETLANDS: DEFINITIONS AND KEY BIOLOGICAL COMPONENTS

Great Basin wetlands are generally characterized by continuous expansion and contraction, both seasonally and over much longer periods of time. This fluctuation, driven primarily by varying precipita-

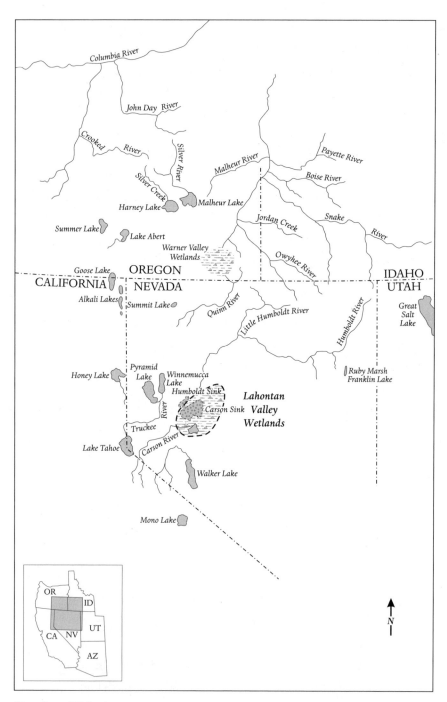

Great Basin Wetland

tion cycles, helps to create a diversity of wetland habitat types within a localized area. It is quite common to find a relatively freshwater wetland at the inflow gradually transforming into more saline wetlands farther down the system. In addition, within the span of one season, these wetlands can transform from fresh, clear lakes to brackish marshes with extremely high salt concentrations (USFWS 1996). Deep perennial marsh is uncommon in the Great Basin; more typical is a wetland system that undergoes periodic, if not annual, drying. The invertebrate and plant productivity of Great Basin wetlands is tied to these boom-and-bust cycles. Dry periods allow the soils to aerate and give winds the opportunity to remove salts that have built up in the soils.

The salt flat, or playa, typically constitutes the bottom end of Great Basin wetlands and also is what remains at the end of a wet cycle or season once evaporation has dried up the wetland. Playas are highly productive wetlands, and are particularly important to many shorebird and some duck species.

LAHONTAN VALLEY WETLANDS: A GREAT BASIN OASIS

The Lahontan Valley wetlands on the western edge of Nevada are part of a network of lakes and wetlands sprinkled across the Great Basin. These aquatic systems are Pleistocene remnants that provide important migratory and breeding habitat for millions of birds. A number of these systems, including the Great Salt Lake, Mono Lake, and the Lahontan Valley wetlands, have received recognition as part of the Western Hemispheric Shorebird Reserve Network.

In addition, researchers and birders are learning more about other Great Basin sites, including Franklin Lake and Ruby Marsh in northeastern Nevada, the Warner Basin wetlands in Oregon, Abert and

Summer Lakes in Oregon, the Humboldt Sink, as well as Pyramid and Walker Lakes in western Nevada and the Alkali Lakes and Honey Lake in California (see map p. 20). Unfortunately, a number of these lakes and wetlands are threatened by human activities. We are just beginning to understand how these areas are interconnected as breeding and migratory sites, and more research and observation are vital if we are to understand how birds use these sites and then make a strong case to protect what remains or restore what has been degraded or lost.

NATURALISTS AND ORNITHOLOGISTS IN THE LAHONTAN VALLEY

Our understanding of the Lahontan Valley and its bird life has benefited from investigations by professional and amateur biologists over the past hundred years. The first field notes from the Lahontan Valley were recorded by Vernon Bailey and H. C. Oberholser during their May 2–10, 1898, trip from Ragtown to Stillwater. At the time, Bailey was the chief naturalist of the U.S. Biological Survey, a position he held from 1887 to 1933. His interest in the Lahontan Valley was prompted largely by the presence there of his sister, Anna Bailey Mills, who was married to a local farmer and whom Vernon came to visit periodically.

Although he never formerly trained as a biologist, Anna Bailey Mills's son, Vernon Mills, grew up in the Lahontan Valley with his mother and uncle encouraging his interest in natural history. A lifelong observer of birds in the valley, Mills never systematically recorded his observations in field notes, but many of his observations were captured opportunistically by his friend Ray Alcorn. Mills's sightings became an important part of Ray Alcorn's *The Birds of Nevada*.

The most consistent observer of the Lahontan Valley's bird life was J. R. (Ray) Alcorn, who moved to Fallon with his family in 1929. He struck up a friendship with local teenager Vernon Mills, and his natural history interests blossomed under the tutelage of Anna Bailey Mills, who was an accomplished naturalist in her own right. In 1936, Alcorn took a job in Fallon in the Predator-Rodent Control Division of the U.S. Biological Survey, and a long and multifaceted biological career was launched. Alcorn eventually left Fallon to assist Dr. Raymond Hall as a field biologist. He collected specimens for Hall throughout North and South America from 1947 to 1960, then returned to Fallon as the USFWS's district supervisor of the Predator-Rodent Control Division. After he retired from the USFWS in 1973, Alcorn busied himself with compiling the Nevada bird records that had accumulated since the publication of Jean Linsdale's "The Birds of Nevada" in 1936. In 1988 he published *The Birds of Nevada*.

Our knowledge of the valley's bird life benefited greatly from Linsdale's visits in the early 1930s as part of the fieldwork he undertook prior to writing "The Birds of Nevada," the first compilation of Nevada's bird records.

Since the 1950s, a series of federal and state biologists have provided extremely detailed accounts of the valley's wetlands, birds, and other wildlife. Dave Marshall, USFWS biologist for the Stillwater Wildlife Management Area (WMA) from 1950 to 1954, was instrumental in establishing a prereclamation picture of the wildlife resources of the Lahontan Valley through his extensive discussions with Wuzzie George and Alice Steve, two local Native American women with knowledge of traditional subsistence methods. Larry Napier, USFWS biologist for the Stillwater WMA from 1967 to 1974, provided the first comprehensive population and production estimates for non-game-bird species. Napier's work is representative of the broadening interest in the Lahontan Valley's full wildlife diversity during those years. Steve Thompson, USFWS biologist at Stillwater WMA from 1985 to 1989, in collaboration with Larry Neel, established monitoring programs for shorebirds and colony-nesting birds that are still performed today and provide much of the baseline data presented in this book. On his arrival at Stillwater, Thompson found himself embroiled in the last-ditch effort to save the Lahontan Valley wetlands, and he did much to publicize the importance of the wetlands in broader scientific and conservation circles. His work contributed to the inclusion of the Lahontan Valley in the Western Hemispheric Shorebird Reserve Network as a "site of hemispheric concern" in 1988, as well as the nomination of the Lahontan Valley to the Ramsar Delegation of Wetlands of Global Concern, which is still under consideration. Finally, Norm Saake, Nevada Division of Wildlife's waterfowl biologist since 1967, established himself as the Lahontan Valley wetlands' most effective and tenacious advocate. Saake's tireless energy and single-minded dedication have left an indelible mark on all aspects of the wetlands' biology and conservation—from survey and inventory, wetlands management and water modeling, to coalition building and policy formation at the local, state, and national levels. The protection of the Lahontan Valley wetlands has been his foremost concern.

Our knowledge of the Lahontan Valley wetlands has also been enriched by the knowledge and insights of Wuzzie George, one of the last Fallon Paiutes born in the Stillwater wetlands. She spent much of her childhood with her grandparents learning about the vanishing subsistence lifestyle of her people. Wuzzie George can truly be called an influential Lahontan Valley ornithologist of a very natural order. Her knowledge of the breeding and migratory status of almost all the native bird species amazed biologists such as Dave Marshall and Ray Alcorn, who grilled her with questions about study skins. Her knowledge and reliability gave biologists insight into the nature of the valley's prereclamation wildlife as well as an understanding of the wetlands prior to the large-scale changes brought about by irrigation and management.

The increasing popularity of birdwatching in the past decade has provided additional information on the valley's bird life, particularly on nongame species.

CHANGING SEASONS

Winter

As the winter freeze begins locking up the wetlands, all but the hardiest migrants are pushed out of the Lahontan Valley. The cold weather brings northern birds south, including bald eagles, rough-legged hawks, and northern shrikes. During hard freezes the birds congregate at any source of open water, especially agricultural drains. Cliff-nesting birds such as prairie falcons and golden eagles also move into the valley during the winter. During mild winters it is not unusual to find ibis, egrets, and herons as well as a small number of shorebirds trying to overwinter.

Spring

Tens of thousands of waterfowl, including snow geese, gadwalls, northern pintails, and green-winged and cinnamon teal, begin returning in February and early March. American white pelicans also start returning by late February. Many of the winter raptors, including bald eagles, northern harriers, rough-legged hawks, and short-eared owls, remain into late March and early April. Nesting ducks begin breeding in late March and April. Shorebirds begin arriving in small numbers in March. Migration peaks in the third week of April, bringing thousands of American avocets, black-necked stilts, long-billed dowitchers, western and least sandpipers, long-billed curlews, and many other species. One or more peregrine falcons usually visit the valley to take advantage of the growing concentration of shorebirds. Land birds mostly begin arriving in April, peaking in early May, when residents such as house and Bewick's wrens, lazuli buntings, black-headed grosbeaks, and Bullock's orioles begin breeding. In early June, the last of the migrants, including common nighthawks and flycatchers, pass through.

Summer

By early May, the large number of colony nesters, including white-faced ibis; great, snowy, and cattle egrets; great blue herons; and black-crowned night herons, have reoccupied their colonies. In years of plentiful water, Franklin's gulls and Forster's and black terns will also nest in the valley. In late summer, large numbers of American white pelicans congregate throughout the valley wherever fish are available. Meanwhile, the marshes are alive with the sights and sounds of many other species, including American bitterns, Virginia and sora rails, marsh wrens, and yellow-headed blackbirds.

Fall

Fall migration begins early with Wilson's phalaropes beginning their migration in late June. By August, the thousands of phalaropes are followed by dowitchers, sandpipers, yellowlegs, and other shorebirds returning south. The first cold fronts in late August push large groups of avocets and stilts out of the valley. Land bird migrants, including flycatchers, vireos, warblers, and tanagers, start moving through the Carson River corridor beginning in the middle of August and continuing through the end of September. Beginning in September, wave after wave of waterfowl arrive. Cold weather in October pushes large numbers of white-crowned sparrows, red-breasted nuthatches, mountain chickadees, and other montane species down into the valleys. Hermit thrushes begin showing up in November.

THE DATABASE

The federal and state biologists who have worked in the Lahontan Valley have generally been more concerned with the valley's wetland habitat than with the drier areas. This understandable interest in the unique wetlands means that much of the knowledge about the valley's avifauna is focused on species associated with the wetlands, and on waterfowl in particular. More recently, colony-nesting species such as ibis, egrets, and herons, as well as shorebirds, have been increasingly studied and surveyed.

The traditional focus on the wetlands and their species is apparent in the species accounts that follow this chapter: quite simply, we know more about the status, seasonal occurrence, and in some case long-term trends for these species. Our knowledge of the valley's songbirds, particularly migrants, is less complete. In places, we supplement the valley data with information on broader trends in the Truckee-Carson-Walker basins. We hope that readers of this book will take the opportunity to add to the knowledge of the valley's avifauna by spending more time along the Carson River and adjoining upland sites. We suspect that many of our Neotropical migrants use low-elevation riparian sites such as the lower Carson River during migration and that these sites play an important role in the long-term health of these species.

Nomenclature

We follow the nomenclature and order of the seventh edition of the American Ornithologists' Union's *Checklist of North American Birds* (1998).

Criteria for Acceptance of Records

As a rule, we have accepted records that generally conform with a particular bird's pattern of occurrence in western Nevada. We have tried to be somewhat more critical with rare and out-of-season sightings. In these cases we have relied on a combination of written and verbal descriptions, specimens, photographs, and observer expertise. The Nevada Bird Records Committee was formed only recently and has reviewed relatively few of the records. The committee has accepted some reports and rejected others due to lack of documentation.

Status Terminology

COMMON: Always or almost always encountered in the appropriate habitat, sometimes in large numbers, without special searching.

UNCOMMON: Encountered in small numbers, and sometimes in large numbers, but usually missed unless a special search is made in the appropriate habitat.

RARE: Cannot be expected on any given day or even every year.

VAGRANT: Fewer than 5 records.

SOURCES

We have done our best to track down and comb through all the available field notes, publications, databases, and reports, as well as interviewing knowledgeable individuals. The Stillwater National Wildlife Refuge and the Nevada Division of Wildlife remain important repositories of information regarding the valley's bird life. We made a special effort to locate specimens housed in museums, including the Barrick Museum of Natural History at the University of Nevada, Las Vegas; the Museum of Vertebrate Zoology at the University of California, Berkeley; the Museum of Vertebrate Zoology at the University of Nevada, Reno; the Natural History Museum at the University of Kansas; and the U.S. Museum of Natural History. In addition, we have drawn on reports in *Audubon Field Notes* and *American Birds,* published by the National Audubon Society. We list additional sources in the bibliography.

List of Abbreviations and Observers

GA	=	Gene Albanese	HJ	=	Hugh Judd
KB	=	Keri Boomgarden	JJ	=	Joe Jehl
LB	=	Larry Bowman	PL	=	Paul Lehman
RB	=	Rob Bundy	DM	=	David McNinch
GC	=	Graham Chisholm	MM	=	Martin Meyers
BEC	=	Beth Clark	VM	=	Vernon Mills
BC	=	Bill Clark	MN	=	Martha Neel
LC	=	Luke Cole	LN	=	Larry Neel
GCO	=	Gary Cottle	RN	=	Ray Nelson
JD	=	Jon Dunn	JP	=	Jim Paruk
JE	=	Jim Eidel	NS	=	Norm Saake
CE	=	Chris Elphick	GS	=	Greg Scyphers
RF	=	Robert Flores	JS	=	Jeff Sealy
TF	=	Ted Floyd	DS	=	Dennis Serdehely
KG	=	Keith Geluso	RS	=	Rich Stallcup
RH	=	Rich Heap	BS	=	Brian Sullivan
WH	=	William Henry	GT	=	Greg Tanner
CH	=	Chris Hofer	JT	=	Joanne Tanner
MI	=	Marshall Iliff	ST	=	Steve Thompson
AJ	=	Anne Janik	DT	=	Dennis Trousdale

JW	=	Jack Walters	NWR	=	National Wildlife Refuge
DW	=	Diane Wong	USFWS	=	U.S. Fish and Wildlife Service
AB	=	*American Birds*	WMA	=	Wildlife Management Area
CBC	=	Christmas bird count			
FN	=	*Field Notes*			
MOB	=	multiple observers			
NAB	=	*North American Birds*			
NBRC	=	Nevada Bird Records Committee			

Museums

MVZ CA	Museum of Vertebrate Zoology at the University of California, Berkeley
MVZ NV	Museum of Vertebrate Zoology at the University of Nevada, Reno
BMNH	Barrick Museum of Natural History at the University of Nevada, Las Vegas
NHM	Natural History Museum at the University of Kansas
USNM	U.S. National Museum of Natural History

NOTES

1. This narrative was written by Larry Neel and adapted from Donald K. Grayson, *The Desert's Past: A Natural Prehistory of the Great Basin* (Washington D.C.: Smithsonian Institution Press, 1993).

2. This narrative was written by Larry Neel and adapted from Grayson, *The Desert's Past.*

2.
Species
Accounts

FAMILY GAVIIDAE: LOONS

Pacific Loon
(*Gavia pacifica*)
STATUS: Very rare migrant.

The 5 valley records include a specimen collected on April 9, 1950, at Soda Lakes (USNM no. 491175), and 4 sight records: April 24, 1998, at Big Soda Lake; May 3, 1994, at Big Soda Lake; October 26, 1997, at Little Soda Lake; and November 18, 1994, at Little Soda Lake (AB 49:74). Pacific loons are rare but regular migrants at nearby Pyramid and Walker Lakes.

Common Loon
(*Gavia immer*)
STATUS: Rare but regular migrant.

Common loons occur in the Lahontan Valley from mid-April (April 16, 1992, earliest record) through May (May 31, 1975, latest record), and from October through November (November 24, 1988, latest record). Soda Lakes, Lahontan Reservoir, and Stillwater NWR are the most likely locations, but on occasion this species occurs in large irrigation canals. There are 3 July records: July 12, 1992, at Big Soda Lake; July 21, 1993, at Lahontan Reservoir; and July 30, 1990, at Big Soda Lake. There are 2 migratory common loon staging areas in the region: at Walker Lake, south of the Lahontan Valley, and Pyramid Lake, north of the Lahontan Valley. High count: 12 loons, at Big Soda Lake, November 12, 1994.

Pied-billed Grebe

(*Podilymbus podiceps*)

STATUS: Common summer resident and migrant; uncommon winter resident.

This species usually arrives in April (March 25, 1987, earliest record) and leaves by November 1. Small numbers will overwinter if open water remains. Pied-billed grebes occur regularly at Stillwater NWR, Carson Lake, and S-Line and Harmon Reservoirs. They seem to be particularly fond of the thick mats of aquatic buttercup (*Ranunculus* sp.) that flourish in fresher waters, particularly in Indian Lakes and S-Line Reservoir. On September 1, 1995, hatching eggs were observed in a nest at Carson Lake, an extremely late date.

Horned Grebe

(*Podiceps auritus*)

STATUS: Rare but regular spring and fall migrant.

Spring migrants have been recorded from the end of February (February 23, 1995, earliest record) through early May (May 10, 1997, latest record). Fall migrants occur from the end of August (August 5, 1997, earliest record) through mid-November (November 14, 1994, latest record). Horned grebes usually occur singly in congregations of eared grebes at the Soda Lakes and Lahontan Reservoir.

Red-necked Grebe

(*Podiceps grisegena*)

STATUS: Vagrant.

There is a single sight record of an alternate-breeding-plumage bird at Harmon Reservoir, August 23–25, 1994 (AB 48:966). This species is an annual rare fall migrant at Pyramid Lake.

Eared Grebe

Eared Grebe

(*Podiceps nigricollis*)

STATUS: Common summer resident and migrant;
uncommon winter resident.

Eared grebes can be found all year in the Lahontan Valley, except in
the coldest winters. Numbers of spring migrants peak in April and
May with counts as high as 10,000. Fall migration peaks in August
and September. During migration, Big Soda Lake is the preferred
site. Eared grebes commonly nest at Carson Lake (310 nests in 1995),
Stillwater NWR (450 nests in 1995), and the regulatory reservoirs if
water conditions are right (58 nests in Harmon Reservoir in 1994).
Nests are constructed on thick beds of pondweed or wigeon grass
that have reached the water's surface.

Western Grebe

Western Grebe

(*Aechmophorus occidentalis*)

STATUS: Common summer resident and migrant; rare winter resident.

Western grebes usually appear in the Lahontan Valley in late February (January 19, 1996, earliest record) and depart in mid-October (December 9, 1957, latest record). They nest at Stillwater NWR, Carson Lake, Harmon Reservoir, and occasionally at other wetlands when habitat is available.

Clark's Grebe

Clark's Grebe

(*Aechmophorus clarkii*)

STATUS: Common summer resident and migrant; rare winter resident.

Clark's grebe is by far the more common of the 2 *Aechmophorus* grebes in Lahontan Valley. This species usually appears in early April (March 6, 1995, earliest record) and departs in October (November 10, 1994, latest record). They nest at Stillwater NWR, Carson Lake, Harmon Reservoir, and occasionally at other wetlands when habitat is available.

American White Pelican

(*Pelecanus erythrorhynchos*)

STATUS: Common resident and migrant.

White pelicans generally arrive in early March (February 21, 2000, earliest record). The Pyramid Lake nesting colony at Anaho Island normally initiates breeding in April, although in 1987 USFWS biologists estimated that eggs were laid as early as February. Anaho Island pelicans regularly visit the Lahontan Valley during the nesting season— up to 1000 pelicans a day during March through June. In addition, nonbreeding birds summer in the valley. Postbreeding concentrations as large as 3000–8000 individuals have been counted, with small numbers remaining as late as December. During unusual climatic and water conditions in 1986–87, 20,000 pelicans overwintered at the Carson Sink.

American White Pelican

There is evidence of ancient nesting colonies from Hazen and Rattlesnake Hill in Fallon. During periods of heavy runoff from the Carson River into the Carson Sink, white pelicans have been confirmed as nesters (e.g., June 3, 1953 [110 nests], June 17, 1986, and July 21, 1999 [20 nests]).

Brown Pelican
(*Pelecanus occidentalis*)
STATUS: Vagrant.

Jean Linsdale (1951) reported a sight record of a brown pelican flying 10 miles north of the town of Stillwater on May 20, 1934. This is the only northern Nevada record, although there are numerous records from Lake Mead and a record from Kirch WMA.

FAMILY PHALACROCORACIDAE: CORMORANTS

Double-crested Cormorant
(*Phalacrocorax auritus*)
STATUS: Common summer resident and migrant;
rare winter resident.

This species usually arrives in March (February 23, 1996, earliest record) and remains until early November. During the flood years of 1985–87, the valley's breeding population of double-crested cormorants exploded. Colonies were established at S-Line Reservoir, Sheckler Reservoir, and the mouth of the Carson River. In May 1987, 200 cormorant nests were counted in the valley. By June, the sink had almost dried and the Sheckler and Carson River colonies had been abandoned. In 1988, no cormorant nests were successful. Since that drastic crash, the breeding population has built up again to about 40 pairs, mostly at Gull Island in Lahontan Reservoir.

FAMILY ARDEIDAE:
BITTERNS AND HERONS

American Bittern
(*Botaurus lentiginosus*)
STATUS: Uncommon summer resident and migrant;
rare winter resident.

American bitterns are most common in the valley between April and
November. They overwinter in small numbers at Carson Lake and
Stillwater NWR.

In C. S. Fowler's landmark ethnography, *In the Shadow of Fox Peak*,
the American bittern is identified as *tabapunikadii*, or "sits looking at
the sun." Wuzzie George, the primary source of the book's natural

American Bittern

history observations, considered American bitterns more common when she was a girl (ca. 1900) than they were during her later years (Fowler 1992).

Least Bittern
(*Ixobrychus exilis*)
STATUS: Rare summer resident.

Least bitterns have been recorded from early May (?) through September (October 4, 1950, latest record). This species may be present in extremely small numbers annually, but mostly goes undetected. Carson Lake and Stillwater NWR provide most of the records, but least bitterns have also been recorded at the south end of Harmon Reservoir and Letter Reservoir along the lower Carson River. A recently fledged least bittern found along Beach Road near Carson Lake on July 20, 1986, provided the first evidence of breeding (LN). In addition, an immature was seen on July 26, 1997, at Carson Lake (CE); and an unfledged juvenile was seen on August 17, 1998, at Carson Lake (NS).

Great Blue Heron
(*Ardea herodias*)
STATUS: Common summer resident; uncommon winter resident.

Nesting begins in late February. There are scattered rookeries in the valley, including at Carson Lake, Stillwater NWR, Timber Lakes, the Canvasback Club, and S-Line Reservoir. Heron nesting success appears tied to the availability of habitat that supports a fish population. The number of nesting pairs varies greatly depending on the availability of habitat (e.g., a high of 657 nests in 1986, and a low of 20 nests in 1988).

Great Blue Heron

Great Egret

Great Egret
(*Casmerodius albus*)
STATUS: Common summer resident;
uncommon to rare winter resident.

This species usually arrives in the Lahontan Valley in the first week
in April, and incubation of eggs is under way by May 15. Most leave
the valley by November 1, but some stay through mild winters (40
recorded in January 1996). Nesting colonies are located at S-Line

Reservoir, Carson Lake, Canvasback Club, Stillwater Point Reservoir, and Lahontan Reservoir. The nesting population fluctuates depending on the availability of habitat (e.g., the high nesting pair count was 485 in 1987; the low count was 80 in 1988).

Snowy Egret

Snowy Egret

(Egretta thula)

STATUS: Common summer resident; rare winter resident.

Snowy egrets arrive in the Lahontan Valley in early April (March 21, 1996, earliest record). Nesting colonies are usually well established by mid-May. Snowy egrets seem somewhat more adaptable to drying marsh conditions than great egrets or great blue herons, and will abandon, relocate, and renest later in the season if necessary. They leave the valley by November 1, and are rarely recorded in winter. Regular breeding colonies occur at Carson Lake, Canvasback Club, S-Line Reservoir, and Gull Island in Lahontan Reservoir. In addition, during very wet years, breeding colonies have started at Sheckler Reservoir and the mouth of the Carson River. Nesting pair counts for this species have varied from a high of 330 in 1987 to 145 in 1989.

Cattle Egret

(Bubulcus ibis)

STATUS: Common summer resident.

Cattle egrets have been recorded in the valley from April (April 4, 1997, earliest record) through October 1. During mild winters they overwinter in small numbers. This species was first recorded in the valley in spring 1977 when 6 birds were at Stillwater NWR (AB 31:1028). Numbers increased in the valley from the time they were first reported nesting at Carson Lake in 1980 until 1990, then leveled off. Cattle egrets have breeding colonies at Carson Lake, Canvasback Club, and S-Line Reservoir. High nesting pair count: 225 in 1990.

Green Heron

(Butorides virescens)

STATUS: Rare late summer visitor.

Green herons occur in the Lahontan Valley from August through October (e.g., August 26, 1994, at Carson Lake; September 10, 1970, at Big Indian Lake; September 19, October 1, 1994, at Harmon Reservoir; and October 16, 1995, at Stillwater NWR). A sighting from the first week of June 1981 at the Nutgrass unit at Stillwater NWR and a report from east of Fernley along the Truckee Canal, just outside the valley, on June 8, 1958, are the earliest summer records.

Black-crowned Night Heron

(Nycticorax nycticorax)

STATUS: Common summer resident; uncommon winter resident.

Black-crowned night herons generally arrive in the valley in mid-April (March 2, 1996, earliest record), and most are gone by the first of November. A few birds will overwinter, roosting in Russian olive thickets along the Carson River corridor and feeding in potholes along the river and in a few persistent drains.

Nesting dynamics of this species are the most difficult to figure of any of the colony-nesting herons. Prior to 1988, May aerial counts were often quite high, followed by a June count often only one-fifth as large. Nesting numbers were heavily affected by the extended drought after 1988 and have been slow to recover. Colonies since 1986: Carson Lake; Canvasback Club; S-Line Reservoir; Stillwater Point Reservoir; Gull Island, Lahontan Reservoir. Colonies at Sheckler Reservoir and the mouth of the Carson River were active in 1986–87. The high nesting pair count was 1800 in 1987; 1520 nests were reported at Stillwater WMA on June 15, 1954. The peak count of 5000, mostly immature birds, was reported in September 1985 at Stillwater NWR.

Black-crowned Night Heron

FAMILY THRESKIORNITHIDAE: IBISES

White-faced Ibis

(*Plegadis chihi*)

STATUS: Common summer resident and migrant; rare winter resident.

Ibis arrive around April 15, and most leave by September 15. A small number will remain in the valley as long as there is open water (e.g., 200 at Carson Lake on December 20, 1995, and 50 at Harmon Reservoir on January 2, 1996).

White-faced Ibis

Flocks of white-faced ibis are one of the Lahontan Valley's most familiar summer sights. They are often seen foraging in flooded agricultural fields or in pastures at Carson Lake and flying back and forth between foraging sites and nesting colonies at Carson Lake, Stillwater NWR, and Harmon Reservoir. They nest colonially in bulrush stands.

The Lahontan Valley is the site of one of the most important white-faced ibis nesting colonies in the Great Basin. The others are at Bear River NWR in Utah and Malheur NWR in eastern Oregon. Wetland managers have established a management goal of at least 3000 pairs (see appendix 1, table 1).

FAMILY CICONIIDAE: STORKS

Wood Stork
(*Mycteria americana*)
STATUS: Vagrant.

There are several records from the Lahontan Valley from the 1930s, including a report of 25 birds from July 25, 1930, and reports from June and July 1935 and 1936 (Alcorn 1988). The apparent lack of recent records may be linked to a decline in the wood stork's population in the lower Colorado River basin (Rosenberg et al. 1991).

FAMILY CATHARTIDAE:
AMERICAN VULTURES

Turkey Vulture

(*Cathartes aura*)

STATUS: Common migrant and summer resident.

Turkey vultures usually arrive in the valley the last week of March (March 19, 1961, and 1994, earliest records) and depart by early October (October 5, 1997, latest fall record). A roost of 30–40 birds developed in the village of Stillwater in the 1990s.

FAMILY ANATIDAE:
SWANS, GEESE, AND DUCKS

Fulvous Whistling-Duck

(*Dendrocygna bicolor*)

STATUS: Vagrant.

Alcorn (1988) reported a specimen from a flock of 20 birds taken 14 miles west of Fallon on November 14, 1940 (NHM no. 28955), and 1 shot at Harmon Reservoir October 19, 1951 (MVZ CA no. 125129). There are no confirmed northern Nevada records since 1960. The decline of this species in the lower Colorado River basin may be linked to the lack of recent sightings in the valley (Rosenberg et al. 1991).

Greater White-fronted Goose

(*Anser albifrons*)

STATUS: Uncommon migrant.

A small group of birds thought to originate in the Bristol Bay region in Alaska appears in late August and early September. Another group appears in October. Generally these geese do not remain in the Lahontan Valley very long. Occasionally 1 or 2 overwinter in conjunction with Canada geese. Spring migrants begin returning in mid-January and have usually left the valley by the end of March. Stragglers may remain longer (e.g., a May record, Alcorn 1946). Usually seen in groups of about 20–50; 69 were seen at Carson Lake on September 13, 1995.

White-fronted geese are most common on wet fields, meadows, and shallow marshes at Carson Lake and Stillwater NWR, but are occasionally encountered in small numbers throughout the valley.

Snow Goose
(*Chen caerulescens*)
STATUS: Common migrant.

Snow geese usually arrive in early October (September 29, 1992, earliest date), and numbers peak in November before the winter freeze forces them out of the valley in December (December 23, 1995, latest date). Radar tracking data indicate that Lahontan Valley birds move to California's Central Valley. Snow geese begin returning in mid-January (January 12, 1995, earliest date), with the last birds moving through in March and very rarely in early April (April 21, 1989, latest date). Flocks are usually larger during spring migration. In 1994, approximately 30,000 birds had gathered at Carson Lake in mid-March. Small numbers of blue-phase snow geese are present annually in large flocks of white geese. Although snow geese occasionally occur throughout the Lahontan Valley, Carson Lake is the preferred site. Collared birds from as far away as Wrangell Island, Alaska, have been identified in the Lahontan Valley flocks.

Ross' Goose

(*Chen rossii*)

STATUS: Uncommon migrant.

Migration patterns in the Lahontan Valley generally mirror those of snow geese, although Ross' geese may arrive slightly later than the first snow geese in late October and early November. A single free-flying bird seen on May 3, 1996, at the Carson River Diversion Dam was an extremely late record. Ross' geese account for about 5 percent of the white geese in Nevada (NS). They intermingle with snow geese flocks, but also occur in small pure flocks (e.g., 36 at Carson Lake on March 14, 1995).

Canada Goose

(*Branta canadensis*)

STATUS: Common resident.

The vast majority of Canada geese in the Lahontan Valley are the large *B. c. moffiti*. These birds are regional residents that shift season-ally between Truckee Meadows, Humboldt Sink, the Walker Basin, and the Lahontan Valley. The 1987–94 drought greatly reduced their numbers, but populations rebounded quickly. Mid-1990s winter counts indicated 4500–5000 birds of this race.

Lesser Canada geese (*B. c. parvipes*), a smaller, darker race, usually arrive after mid-November (November 6, 1991, earliest date) and over-winter in the valley before leaving at the beginning of March (March 10, 1993, latest date); annual totals average around 3000 birds. An occa-sional small flock of cackling geese (*B. c. minima*) is observed in the valley between November and February (April 12, 1997, latest date). During the winter months, Canada geese of both major subspecies roost in the portion of Stillwater NWR closed to hunting and make daily feeding forays into surrounding agricultural fields, particularly in the Stillwater District.

Brant

(Branta bernicla)

STATUS: Rare migrant.

Individual birds and small flocks (e.g., 14 birds at Carson Lake on October 28, 1970) occur occasionally in the Lahontan Valley. There are also records from Pyramid Lake, Walker Lake, Mason Valley, and Truckee Meadows. Fall records cluster in October (e.g., October 4, 1957, at Old River Reservoir; October 25, 1970, at Stillwater NWR; and October 28, 1970, at Carson Lake). Spring records include April 29–30, 1970, at S-Line Reservoir; and May 12, 1970, with no location (Alcorn 1988). There is 1 winter record of a bird shot on January 25, 1969, at Stillwater WMA.

Trumpeter Swan

(Cygnus buccinator)

STATUS: Rare winter visitor.

Apparently this species is a relative newcomer to the valley following its introduction in Ruby Lake NWR in 1955 and other sites in adjoining states. There are almost annual reports, usually of individual birds, in the Lahontan Valley beginning in late November (November 26, 1991, earliest fall record). Several marked birds from Summer Lake, Oregon, have appeared in recent years (N. Saake, pers. comm. 1996).

Tundra Swan

(Cygnus columbianus)

STATUS: Common migrant; irregular winter resident.

The Lahontan Valley is a major staging and wintering site for tundra swans, with peak numbers now ranging from 1200 to 2000 birds. The average peak in the 1949–70 period was 3783, with a high count of

12,700 swans December 15, 1956, at Stillwater NWR. They usually appear in late October (late September 1968, earliest record), and only total freeze-up in late December drives them out of the valley for warmer climes in California's Central Valley. A few stragglers usually remain through the coldest part of the winter. Tundra swans often return in mid-January; their numbers build to a peak in January and early February, and they depart by late March (March 28, 1991, latest record).

Tundra swans are prodigious consumers of submergent marsh vegetation, their favorite being sago pondweed, which grows in the mid-saline shallow units of Stillwater NWR and Carson Lake. The swans use their strength and bulk to maintain open water against the relentless encroachment of ice through the cold winter months, thus performing a beneficial service for ducks and coots, which also need open water for feeding.

Wood Duck

(*Aix sponsa*)

STATUS: Common resident.

Wood ducks are found along the Carson River and in sloughs and drains. The invasion of Russian olive and the installation of wood duck nest boxes along the Carson River have increased numbers. During the 1970s this species was very uncommon. Numbers in the Lahontan Valley today may be 200–300 birds, including concentrations of up to 150 birds occurring on ice-free pools along the old Carson River channels in the Sheckler–St. Clair district during winter months. There appears to be no migratory influx of this species into the valley.

Gadwall

(*Anas strepera*)

STATUS: Common summer resident; common migrant; uncommon winter resident.

Fall migration peaks in late October and early November; most leave in late December or early January. Spring migrants return in late January. This species is increasing in the Lahontan Valley and is now the third most common waterfowl nester, with an annual average of 389 nests from 1967 to 1995. Gadwalls prefer deeper marsh than teal, but also frequent flooded grasses, mudflats, and bulrush stands. High count: 26,135 birds, October 12, 1970 (AB 25:84).

Eurasian Wigeon

(*Anas penelope*)

STATUS: Rare migrant.

Records include February 1, 1997, at Carson Lake; February 25, 1968, at Stillwater NWR; March 28, 1988, at Carson Lake; an October 1993 record from Carson Lake; November 6, 1985, at Carson Lake; and November 14, 1996, at Swan Lake, Stillwater NWR. This species is being found more regularly in northern Nevada, perhaps because there are more observers in the field.

American Wigeon

(*Anas americana*)

STATUS: Rare summer resident; common spring and fall migrant; uncommon winter resident.

Fall migrants begin to arrive in September, and numbers peak in October and November. A few overwinter in harsh winters if open water remains in drains. Spring migrants return in mid to late February and have left by May 1. They are generally absent from the valley from

May 1 through August 1. American wigeon frequent flooded grass and wet meadows. High count: 34,500, October 12, 1970, at Stillwater WMA (AB 25:84).

American Black Duck
(*Anas rubripes*)
STATUS: Vagrant.

A banded adult female was recovered on November 4, 1963, by W. D. Lewis 6 miles west-southwest of Fallon. The bird was banded at Wilson Hill, N.Y., on September 20, 1962 (Alcorn 1988). There is no information as to the status of the specimen.

Mallard
(*Anas platyrhynchos*)
STATUS: Very common year-round resident.

The mallard is a common nester in the valley, with an estimated 2500–3000 young raised annually. Their numbers peak in the fall at 12,000–15,000 birds.

Blue-winged Teal
(*Anas discors*)
STATUS: Rare summer resident; uncommon spring, and perhaps fall, migrant.

Most records are from April to July. In summer and fall eclipse plumage makes identification difficult, although there is a record from September 16, 1996 (RB). There is 1 winter record from December 16, 1963. Usually seen singly or in pairs. A male with seven young at Stillwater NWR on June 29, 2000, confirmed breeding in the valley (WH).

Cinnamon Teal

Cinnamon Teal

(*Anas cyanoptera*)

STATUS: Common summer resident and migrant;
rare winter resident.

Numbers peak in September and drop off very quickly by the end of
October, although Saake reported an influx of about 200 birds at Carson Lake in December (December 30, 1989, latest record). Spring migrants return in late February (January 28, 1988, earliest record), and
numbers peak again in April. The cinnamon teal is one of the 2 most
common waterfowl species nesting in the valley (the redhead is the
other), with an annual average of 995 nests from 1967 to 1995. Cinnamon teal nest throughout the valley, including in agricultural drains.

Northern Shoveler

(*Anas clypeata*)

STATUS: Common migrant; uncommon summer resident; uncommon winter resident except for the coldest winters.

Fall migrants generally appear in the second week of August, and numbers peak in late October and early November. Freezes will move them out of the valley. Very abundant during spring migration, March–April. Nesting occurs at Stillwater NWR an average of one year in five. Shovelers nest more commonly outside the valley at Washoe Lake (Washoe County). High count: 26,270, November 7, 1952, at Stillwater WMA.

Northern Pintail

(*Anas acuta*)

STATUS: Uncommon summer resident; common spring and fall migrant; uncommon winter resident.

Drakes begin arriving from breeding grounds to molt in the second week of July. Fall migration peaks in late October or early November. This species winters in all but the most extreme years in the valley. Spring migration peaks in March, with the majority leaving the valley in early April. High count: 50,500, March 7, 1952, at Stillwater WMA.

Green-winged Teal

(*Anas crecca*)

STATUS: Common spring and fall migrant; very uncommon summer resident.

Spring migration peaks from mid-March through early April. Fall migrants begin to arrive in mid-August and peak in October. Green-winged teal will overwinter in smaller numbers as long as open water remains. Following freezes this species will move into drains where flowing water continues to provide habitat. They are almost completely gone by May 1, with a very few remaining in the summer. Perhaps 12 pairs per year nest in the valley. They typically frequent shallow flooded fields and pastures, and also mudflats when there is a good invertebrate bloom. High count: 90,000, 1964. Recent peaks include 9720 at Stillwater Point Reservoir, Stillwater NWR, on November 7, 1996. A sight record of common teal or the Eurasian race of green-winged teal was reported from Mahala Slough on April 4, 1972 (AB 26:788).

Canvasback
(*Aythya valisineria*)
STATUS: Rare summer resident; common spring and fall migrant; uncommon winter resident.

Canvasbacks are most common in the valley in mid to late October and early November; they leave at freeze-up. Spring migrants return in mid-February and are almost all gone by mid to late April. There are usually 5–6 nests each year in the valley (N. Saake, pers. comm. 1996). Stillwater Marsh is one of the top migration areas in the Pacific

Canvasback

Flyway for canvasbacks; peak counts range from 24,000 to 28,000. Sago pondweed is a favored food. Stillwater NWR and Harmon Reservoir are the best sites to see large flocks.

Redhead

(*Aythya americana*)

STATUS: Common summer resident; common spring and fall migrant; uncommon winter resident.

Fall migration peaks during the last week of September through the first 2 weeks of October. Most leave with winter freezes. Spring migration begins in February and peaks in April. Peak migration counts are up to 30,000. Redheads are one of the valley's top two waterfowl nesters, along with cinnamon teal, with an annual average of 1346 nests from 1967 to 1995. Redheads prefer deeper water than cinnamon teal, surrounded by stands of alkali bulrush and hardstem. Stillwater NWR and Carson Lake are the top nesting areas. Massie and Mahala Sloughs were formerly important nesting areas. Migration concentrations occur at Stillwater NWR, Harmon Reservoir, Carson Lake, and S-Line Reservoir.

Redhead

Ring-necked Duck
(*Aythya collaris*)
STATUS: Uncommon spring and fall migrant;
uncommon winter resident.

Fall migrants appear in late October (August 16, 1996, earliest fall record). Spring migrants appear in January and February. This species typically does not remain in the valley late in the spring (May 14, 1994, latest record). High counts are 500–700, and numbers appear to be increasing in recent years. Ring-necked ducks do not nest in the Lahontan Valley. They are typically found with canvasbacks in open water with submergent vegetation.

Greater Scaup
(*Aythya marila*)
STATUS: Rare but regular migrant.

Greater scaup have been recorded from the valley from October through May. Records include October 1, 1994, at Harmon Reservoir; October 25, 1995, at Harmon Reservoir; November 21, 1995, at Tule Lake, Stillwater NWR; January 1, 1958 (NHM no. 35322), January 2, 1968, and January 10, 1969, in the Fallon area; a single bird on April 13, 1997, at Carson Lake; 8 birds on May 10, 1997, at Carson Lake; and a single bird on July 3, 1999, at Carson Lake (DS). This species occurs more commonly at Pyramid and Walker Lakes.

Lesser Scaup
(*Aythya affinis*)
STATUS: Uncommon spring and fall migrant and winter resident.

This species is more common in spring, usually in April, than fall. Records of 3 males on June 3, 1996, at Stillwater NWR; a single bird on June 13, 1994, at S-Line Reservoir; and a pair on June 14, 1995, at Division Pond, Stillwater NWR are unusual. Lesser scaup are not known to nest in the valley.

King Eider
(*Somateria spectabilis*)
STATUS: Vagrant.

A female shot by a hunter at Carson Lake on December 6, 1998 (NAB 53:187), constitutes the first state record for this species as well as the only record in the intermountain West.

Surf Scoter
(*Melanitta perspicillata*)
STATUS: Rare but regular fall migrant.

Surf scoters have been seen in the valley from October through mid-November. Records include October 6, 1995, at Big Soda Lake; October 19, 1940 (2 shot), at Soda Lakes (MVZ CA nos. 80608, 80609); October 19, 1967 (2 birds shot), at Stillwater WMA; October 22, 1967 (1 shot), at Stillwater WMA; October 27, 1987, on Carson River below Lahontan Dam; November 12, 1940 (USNM no. 367942); and November 15, 1949, at Stillwater WMA. More difficult to explain are 2 August sightings from Stillwater NWR: August 14, 1995, from the South Nutgrass (immature, molting); and August 21, 1995, from East Dry Lake (female) (WH).

White-winged Scoter
(*Melanitta fusca*)
STATUS: Rare but regular late fall migrant.

Records for this species include October 13, 1961, and October 23, 1949, at Stillwater WMA; October 28, 1962, and October 30, 1994, at Stillwater NWR; November 2, 1941, November 3, 1940 (flock of 15), and November 4, 1951, at Stillwater WMA; November 7, 1993, at Big Soda Lake; and November 12, 1940 (NHM no. 28956).

Long-tailed Duck
(*Clangula hyemalis*)
STATUS: Rare fall migrant.

Most records are from late October to early January, including: October 20, 1949, at Carson Lake; November 20, 1968, 4 miles northeast of Fallon; November 20, 1989, at Stillwater NWR; November 22, 1952 (NHM no. 32614), and November 23, 1954, at Stillwater WMA; December 3, 1970, at Carson Lake; December 10, 1994 (3 shot in a flock of 20), and January 8, 1994, at Soda Lakes; and March 30, 1995, at Little Soda Lake. There is also a report of a pair present during the summer of 1976 until August 2 (AB 30:983; AB 31:204).

Bufflehead
(*Bucephala albeola*)
STATUS: Common spring and fall migrant and common winter resident.

Buffleheads occur in the valley from October through the end of April (May 15, 1996, latest spring record), with peak numbers occurring December–February. A record from August 8, 1996, was highly unusual. This species frequents open water at Carson Lake, Stillwater NWR, and S-Line and Harmon Reservoirs.

Common Goldeneye
(*Bucephala clangula*)
STATUS: Uncommon spring and fall migrant and winter resident.

Winter residents are present between November (November 6, 1940, and 1992 earliest records) and early March (May 3, 1996, latest record) with peaks during the colder months. Formerly numbers peaked at 100 birds, but current peaks are substantially lower. A June 20, 1966, record was probably of a crippled bird. Goldeneye are most likely to be encountered on the Carson River just below Lahontan Reservoir, Harmon Reservoir, and Soda Lakes.

Barrow's Goldeneye
(*Bucephala islandica*)
STATUS: Vagrant.

A pair was seen on December 17, 1999, and again on December 15, 2000, at the Fallon Sewage Ponds. Schwabenland reported to Alcorn that "an occasional bird has been sighted at Stillwater Marsh."

Hooded Merganser
(*Lophodytes cucullatus*)
STATUS: Uncommon spring and fall migrant; uncommon winter visitor.

Winter birds have been seen from October (October 18, 1993, earliest record) through March (May 16, 1994, latest record). Hooded mergansers are most often seen in open river channels and drains where open water remains in winter.

Common Merganser
(*Mergus merganser*)
STATUS: Common spring and fall migrant; common winter visitor.

Common mergansers occur from November (October 31, 1994, earliest record) through April, and uncommonly in other months. High counts include 1500 on February 17, 1950, at Stillwater WMA. This species nests at Pyramid Lake and on the upper Humboldt River.

Red-breasted Merganser

(*Mergus serrator*)

STATUS: Rare spring and fall migrant; very rare winter resident.

Spring migrants may be present from the middle of April through the third week of May. There are two summer records from June 11, 1992, and July 12, 1992, at Soda Lakes. Fall migrants first appear in the middle of October. There are winter records from December 1, 1957, and December 19, 1997.

Ruddy Duck

(*Oxyura jamaicensis*)

STATUS: Common summer resident and migrant; uncommon winter resident.

Migration peaks in April and November at a couple of thousand birds. Ruddy ducks are fairly common nesters at Stillwater NWR and Carson Lake. Peak numbers have declined from 40,000 in fall and 55,000 in spring 1966 at Stillwater WMA to 26,000 in spring 1977 (AB 31:1028). Ruddy ducks nest over water and prefer deeper water surrounded by cattails and hardstem.

Ruddy Duck

FAMILY ACCIPITRIDAE:
OSPREY, EAGLES, HAWKS,
AND ALLIES

Osprey
(*Pandion haliaetus*)
STATUS: Rare but regular summer resident; uncommon migrant.

Ospreys usually arrive at the end of March (March 24, 1997, earliest record) and depart by mid-September (September 21, 1994, latest record). The Lahontan Valley is one of 4 sites outside the Lake Tahoe basin where ospreys nest in Nevada. Nesting occurred at S-Line Reservoir from 1989 through 2000, excepting 1993. Regulatory reservoirs and Lahontan Reservoir are favorite sites.

Mississippi Kite
(*Ictinia mississippiensis*)
STATUS: Vagrant.

One sight record from the Fallon Golf Course on May 26, 1988 (LN, RH), is the only northern Nevada record for this species.

Bald Eagle
(*Haliaeetus leucocephalus*)
STATUS: Common winter resident.

The first bald eagles typically arrive in the valley around the middle of November (October 27, 1993, earliest record), and most depart by the middle of March (May 19, 1998, latest record).

During the day, individuals disperse from roosting sites to foraging and loafing sites around the valley. Lahontan Reservoir, Carson Lake, Indian Lakes, S-Line Reservoir, and Soda Lakes are the most consistently visited sites, with Lahontan Reservoir supporting roughly half of the wintering contingent each year. The thick cottonwood grove at

Timber Lake is the primary night roost. Secondary roost sites include the sparse line of cottonwoods along the East Ditch of Stillwater NWR (particularly adjacent to the Foxtail units) and a couple of solitary trees north of the York unit of Carson Lake. Lahontan Reservoir birds roost widely in the many cottonwoods that line the perimeter until night temperatures fall below 10°F, when the birds seem to prefer the communal roosts, particularly at Timber Lake. During these cold periods, some interchange between Lahontan Reservoir and Timber Lake seems to occur, but there is also a poorly documented roost in the gallery cottonwood forest on the Carson River directly above the reservoir high-water line. The Lahontan Valley's baseline winter population seems to be 30–35 birds (1986–92). During the fish kill linked to the recession of the Carson Sink in 1987 following the flood of 1986, 53 and 70 bald eagles were counted on surveys on January 12, 1987, and February 27, 1987, respectively. After repeated winter drawdowns of Lahontan Reservoir decimated fish populations there, and as shallow wetland winter acreage disappeared in 1992–93, the wintering bald eagle contingent dwindled as well to a low of 13 in 1995.

A pair of bald eagles successfully nested and raised young at Lahontan Reservoir 1998–2000, following an unsuccessful effort in 1997. This is the first successful breeding in Nevada reported since the 1860s.

Northern Harrier
(*Circus cyaneus*)
STATUS: Uncommon summer resident; common winter resident.

Winter residents are present in the valley from October through April. This species breeds at Carson Lake, Stillwater NWR, Mahala Slough, and S-Line Reservoir. Large winter concentrations of northern harriers (e.g., up to 20 harriers) have been observed at Carson Lake when hardstem bulrush stands were held dry long enough to allow a buildup of vole populations.

Northern Harrier

Sharp-shinned Hawk
(*Accipiter striatus*)

STATUS: Common migrant and winter resident.

Migration begins in late August (August 10, 1994, earliest record) and lasts through October. Many sharp-shinned hawks are found near human settlements in wintertime. Most are gone by the middle of April (May 17, 1996, latest record).

Cooper's Hawk

(*Accipiter cooperii*)

STATUS: Common migrant and winter resident.

Migration occurs from early September (August 1, 1997, earliest record) through October and again from March through late April (May 8, 1998, latest record). Less numerous in winter than sharp-shinned hawks. Cooper's hawks nest in nearby riparian areas upstream on the Carson River. They are occasionally reported in the neighborhoods around Fallon during the summer and may breed in the valley.

Northern Goshawk

(*Accipiter gentilis*)

STATUS: Rare spring and fall transient; rare winter resident?

Northern goshawks are rarely observed in the valley, but it appears that fall migration occurs in September and October and spring migration occurs in February and March, with single birds also observed in April and May. Alcorn reported in 1988 that at least 1 goshawk had been observed annually for the previous 20 years. A specimen was collected January 24, 1944 (MVZ CA no. 89398), 3 miles west of Fallon.

Red-shouldered Hawk

(*Buteo lineatus*)

STATUS: Rare postbreeding and winter visitant?

Red-shouldered hawks visit the Lahontan Valley between the end of September and February, with some years yielding several observations and other years yielding none. The 1990s saw a number of summer records from the Lahontan Valley (a juvenile on August 8, 1993, and a subadult on August 20, 1997) and surrounding areas in the Carson, Truckee, and Walker River basins. It has been suggested that a range expansion may be occurring and that this species may be breeding in western Nevada. Three birds—1 adult and 2 immatures—were seen on December 19, 1997.

Swainson's Hawk

Swainson's Hawk

(*Buteo swainsoni*)

STATUS: Common summer resident.

Swainson's hawks generally return to the valley by late March and early April (March 20, 1994, earliest record). They nest throughout the valley before departing in mid to late August and early September (September 19, 1995, latest record) and are regularly seen perched in cottonwoods and on utility poles.

Alcorn (1988) attributed a decline in Swainson's hawks since 1946 to low numbers of Townsend's ground squirrels. Nesting pairs have increased since 1986 (LN). No regular census of this species has been conducted in the valley.

Red-tailed Hawk

(*Buteo jamaicensis*)

STATUS: Common year-round resident and breeder.

Red-tailed hawks are the most conspicuous raptors in the Lahontan Valley, having adapted to the valley's agricultural landscape. They start nesting in late February.

Ferruginous Hawk

(*Buteo regalis*)

STATUS: Uncommon migrant and winter resident.

Migrating and wintering ferruginous hawks arrive in October (September 13, 1995, earliest record) and depart by February (Alcorn [1988] reported a March observation with no date). A few winter in the valley. Ferruginous hawks seem to favor the more sparsely developed districts, particularly Indian Lakes, Soda Lakes, and the Stillwater area.

Rough-legged Hawk

Rough-legged Hawk

(*Buteo lagopus*)

STATUS: Uncommon winter resident.

Rough-legged hawks generally arrive in early November (October 29, 1996, earliest record) and depart by the end of March (April 16, 1993, latest record). This species can be found throughout the valley, but particularly in the agricultural areas and in the wetlands. A dead female found at Carson Lake on July 21, 1998 (MI, BS), constitutes a rare summer sighting of this Arctic species.

Golden Eagle

Golden Eagle

(*Aquila chrysaetos*)

STATUS: Uncommon year-round resident.

Golden eagles nest on the broken cliffs of the mountain ranges that surround the Lahontan Valley. Three active nests were reported from the valley in 1987 (ST). Alcorn (1988) reported a nest from Stillwater in 1936, and a female incubated through term without hatching young on a nest in a cottonwood tree at Timber Lake in 1986.

FAMILY FALCONIDAE: FALCONS

American Kestrel

(*Falco sparverius*)

STATUS: Common year-round resident and breeder.

The kestrel is a familiar sight in the Lahontan Valley. CBC totals fluctuate from a low of 27 birds in 1992 to a high of 60 in 1994. Kestrels nest in the cavities of old, heart-rotten cottonwoods along the river corridor, delivery ditches, windbreaks, or scattered singly across fields and pastures.

Merlin

(*Falco columbarius*)

STATUS: Rare but regular winter resident and migrant.

Single merlins generally arrive by late October (October 13, 1996, earliest record). Most sightings are probably of migrants that do not remain in the valley for any length of time; however, it appears that 1 or 2 may overwinter. Passage birds can be observed following shorebird flocks into the middle of April (April 16, 1987, latest record).

Alcorn (1988) reported specimens of the *bendirei* race from November 12, 1931, and April 11, 1944, and a specimen of the *suckleyi* race from January 19, 1941. In addition, a *richardsoni*-race merlin was reported on November 26, 1987 (PL).

Peregrine Falcon

(*Falco peregrinus*)

STATUS: Rare but regular migrant; rare winter visitant.

Spring migrants are usually sighted from mid-April through mid-May (March 26, 1990, earliest spring record; May 16, 1988, latest spring record). Fall migrants begin returning in late July (July 8, 1998,

Peregrine Falcon

earliest fall record) and continue passing through into October (October 17, 1994, latest fall record). Usually 1–3 individuals are sighted during spring and fall migrations.

Winter records include sightings on December 30, 1953, at Big Indian Lake; December 18, 1997, at Stillwater NWR; January 11, 1990, at Indian Lakes; and January 30, 1994, at Carson Lake. Peregrines are most often seen at Carson Lake and Stillwater NWR, particularly during peak shorebird migration.

A record from June 24, 1949, at the Canvasback Club within Stillwater WMA raised the possibility that this species might be nesting nearby. In addition, the regular appearance of a single adult peregrine falcon in April and May from 1988 to 1990 raised hopes that there might be a nest in the nearby mountains. Several helicopter surveys of the White Thrones, Desert Range, and Dead Camel Mountains failed to find an eyrie.

Prairie Falcon
(*Falco mexicanus*)
STATUS: Uncommon year-round resident and breeder; perhaps more common in fall and winter.

Prairie falcons are fairly common nesters in the broken cliffs of the mountains that surround the Lahontan Valley. They move into the agricultural areas of the valley to winter. Singles patrolling Carson Lake harass migratory shorebird flocks with limited success. Alcorn (1988) described a reduction in the population of these birds in this area since 1946.

FAMILY PHASIANIDAE:
TURKEYS AND PHEASANTS

Chukar
(*Alectoris chukar*)
STATUS: Nonnative; irregular transient.

Presently, chukar seem to come down into the Lahontan Valley only during the harshest droughts. A small but consistent population persists in the Stillwater Range. Alcorn (1988) noted that chukars were resident but not abundant in the cultivated areas of the valley in 1946, but that they have since disappeared from the cultivated areas. A small number were present at Timber Lakes in August 2000 (DW).

Ring-necked Pheasant
(*Phasianus colchicus*)
STATUS: Nonnative; uncommon and local resident and breeder.

Ring-necked pheasants were first recorded in the valley on April 3, 1941 (Alcorn 1988). Their decline in the agricultural valleys of Nevada since the 1960s is often attributed to a change in agricultural practices that encouraged an earlier first cutting of alfalfa, which occurs during the pheasant's peak nesting period. The decline may also be linked to the indiscriminate importation and stocking of uncertified pen-reared birds, which may spread gallinaceous diseases to established wild populations.

Wild Turkey
(*Meleagris gallopavo*)
STATUS: Nonnative; uncommon year-round local resident and breeder.

Wild turkeys (Rio Grande subspecies) were introduced at several sites along the Carson River in the Lahontan Valley in 1989 and presently maintain a small population in limited habitat. They are often seen from Pioneer Road along the Carson River and along Stillwater Slough near Stillwater Road.

FAMILY ODONTOPHORIDAE: QUAILS

California Quail
(*Callipepla californica*)
STATUS: Nonnative; common year-round resident and breeder.

This species is well established throughout the valley, particularly in agricultural areas. Alcorn (1988) speculated that California quail may be native to the valley based on a conversation he had with Wuzzie George, a local Paiute woman, although he also noted that there were introductions in the 1870s–80s. Fowler (1992), however, claimed that while Wuzzie George was aware of mountain quail from the Sweetwater Mountains, and possibly the Stillwater Range, she made no reference to California quail in her discussions with Fowler of game birds in the Lahontan Valley.

FAMILY RALLIDAE:
RAILS, GALLINULES, AND COOTS

Virginia Rail
(*Rallus limicola*)
STATUS: Common summer resident and migrant; uncommon winter resident.

Birds are generally on their territories by mid-April (March 21, 1996, earliest record); most are gone from the valley by the beginning of

Virginia Rail

November. During harsh winters Virginia rails appear to leave the valley altogether.

Sora
(Porzana carolina)
STATUS: Uncommon summer resident and migrant;
rare winter resident.

Soras arrive in late April (March 26, 1996, earliest record), and most have departed by late October. Their winter status is not well understood, and this species may be overlooked during that season. There are several records from late winter, including February 22, 1996. Generally soras are noted by their horse-whinny calls in the tules.

Common Moorhen

(Gallinula chloropus)

STATUS: Uncommon summer resident; rare winter resident.

Recorded 8 of 13 years on the Fallon CBC with a high count of 10 in 1988. Confirmed breeding records include August 1, 1992, at the Carson River Diversion Dam; and September 1, 1995, at Old River Reservoir. Common moorhens are secretive and local. Although they may be found in any of the tule-cattail marshes around the valley, possibly the most consistent places to find them are in the various tule-choked drainage ditches in the irrigation network.

American Coot

(Fulica americana)

STATUS: Extremely common summer resident and migrant; common winter resident.

In late summer and early fall, coots congregate in large flocks in the wetlands. High counts: up to 200,000 in fall and 60,000 in spring.

American Coot

FAMILY GRUIDAE: CRANES

Sandhill Crane
(*Grus canadensis*)
STATUS: Rare but regular migrant; rare winter visitant.

Most commonly observed in spring, especially in February (May 2, 1898, latest spring record) and October (September 20, 1989, earliest record) through November (December 4, 1995, latest fall record), when small flocks occur in the valley. Most sightings involve flocks of less than 10 birds. A high count of 40 observed from October 4 through November 7, 1995, was unusual. On occasion cranes will overwinter in small numbers, as they did in the winters of 1988–89, 1989–90, and 1993–94. Observers have reported both "lesser" sandhill cranes (*G. c. canadensis*) and a "greater" sandhill crane (*G. c. tabida*). The latter breeds in nearby areas in Douglas County, Nevada; and Sierra, Modoc, Lassen, including Honey Lake, and Shasta Counties in California (LN). Sandhill cranes are most often observed feeding in agricultural fields either near Carson Lake or Stillwater NWR, where they also roost.

FAMILY CHARADRIIDAE: PLOVERS

Black-bellied Plover
(*Pluvialis squatarola*)
STATUS: Uncommon migrant.

Black-bellied plovers are usually seen from early April (March 4, 2000, earliest spring record) to the first week of May (May 15, 1993, latest spring record). Fall migration is unspectacular—individuals and small groups straggle through from the middle of August (August 10, 1988, earliest fall record) through the middle of October (November 19,

Black-bellied Plover

1987, latest fall record). Very unseasonal was a bird seen on December 19, 1993, at Carson Lake. High counts include unspecified April and May records of flocks varying in size from 200 to 400 birds (Alcorn 1988), and 316 birds on April 21, 1989, at Carson Lake (see appendix 1, tables 2 and 3).

American Golden-Plover
(*Pluvialis dominica*)
STATUS: Rare migrant.

Golden-plovers are probable overlooked migrants in the Lahontan Valley, occurring in very small numbers during May (April 26 earliest record) and again in fall from late June (June 21, 1997, earliest fall record) through October (October 22, 1997, latest fall record). Four birds seen on April 10, 1992, at Papoose Lakes, Stillwater WMA, were not identified as either *dominica*- or *fulva*-race birds (WH).

Pacific Golden-Plover
(*Pluvialis fulva*)
STATUS: Vagrant.

There is 1 sight record of an adult on August 11, 1988, at Carson Lake (PL, LN). One bird was photographed on May 17–19, 1996, at Big Soda Lake (AB 50:308). The latter record has been accepted by the NBRC.

Snowy Plover
(*Charadrius alexandrinus*)
STATUS: Uncommon summer resident and migrant.

This species arrives in early April (March 27, 1961, earliest record) and generally departs by the end of September (October 10, 1986, latest record). High counts: 671, 1980; 650, 1987; dropping to 342 (1988) during the 1987–93 drought (Page et al. 1989).

Semipalmated Plover
(*Charadrius semipalmatus*)
STATUS: Uncommon migrant.

This plover generally arrives in the middle of April (March 29, 1991, earliest spring record), numbers peak in the last week of April, and they are gone by the second week of May (May 13, 1988, latest spring record). Fall migrants return in late July (July 18, 1996, earliest fall record) and are gone by the beginning of October (November 10, 1997, latest fall record). They are often found singly or in small groups, often mixed with snowy plovers, along the muddy margins of the watered playas of Carson Lake and Stillwater NWR. The high

Semipalmated Plover

count for a single flock of 76 birds on April 26, 1990, at Carson Lake was unusual; 680 individuals were counted altogether April 21–23, 1989, in the Lahontan Valley.

Killdeer
(*Charadrius vociferus*)
STATUS: Very common summer resident and migrant; common winter resident.

Killdeer are highly adaptable plovers found in wetlands, agricultural fields, and vacant lots. High count: 1800, August 13, 1968.

Mountain Plover
(*Charadrius montanus*)
STATUS: Rare but regular late fall migrant.

The Lahontan Valley is probably visited by this species annually in the winter, but migrants are solitary and easily overlooked. In the valley this species is typically found on grazed pastures at Carson Lake. Individuals do not seem to stay in the valley for any length of time. Most records cluster in November and December. Two birds at Stillwater NWR on October 4, 1971, are the earliest seen. Records vary from 16 birds to a flock of 50 seen on November 6, 1940 (Alcorn 1941a).

FAMILY RECURVIROSTRIDAE: STILTS AND AVOCETS

Black-necked Stilt
(*Himantopus mexicanus*)
STATUS: Common breeder and migrant.

Stilts begin arriving in the valley at the end of March (March 9, 1995, earliest record), and most have departed by the end of August (November 23, 1999, latest record). Spring migration peaks in the middle

Black-necked Stilt

of April and fall migration peaks in mid-August. Stilts are less common than American avocets. High count: 6000 pairs and 2000 young, July 25, 1968, at Stillwater NWR. Black-necked stilts favor areas with more vegetation than avocets prefer.

American Avocet

(*Recurvirostra americana*)

STATUS: Common breeder and migrant.

Avocets arrive in late February (February 2, 1992, earliest record) and remain until November (December 18, 1994, latest record). Peak numbers occur between April 15 and May 1, and between August 1 and August 20. High counts: 51,000 at Stillwater NWR, July 28, 1987; 24,451, August 20–22, 1998, in the Lahontan Valley; and 21,000, August 18, 1988,

American Avocet

at Carson Lake. During mild winters avocets overwinter in the valley (e.g., more than 200 remained at Carson Lake during the winter of 1995–96). Large numbers breed in the valley from early May through the end of July. Breeding pair peaks: 5000+ in 1997 at both Stillwater NWR and Carson Lake. Avocets will also breed in other areas if habitat is available. They prefer open water with less vegetation than black-necked stilts favor.

Margaret Rubega found cows, ravens, harriers, and coyotes among the predators at breeding colonies at Carson Lake (Rubega 1997).

FAMILY SCOLOPACIDAE: SANDPIPERS, PHALAROPES, AND ALLIES

Greater Yellowlegs
(*Tringa melanoleuca*)
STATUS: Common migrant; regular winter resident in small numbers.

Early spring migrants, greater yellowlegs begin arriving at the end of February. Migration peaks in early to mid-April (March 10, 1998, high count of 200 birds), and most are gone by early May (May 15, 1999, latest spring record). Fall migrants arrive in early July (July 1, 1992, earliest fall record) and most leave by October, although it is not uncommon to have 20–30 birds overwintering, particularly in mild winters. This species was recorded on 11 of 13 Fallon CBCs from 1985 to 1997.

Lesser Yellowlegs
(*Tringa flavipes*)
STATUS: Uncommon spring and fall migrant; rare winter visitor.

Generally not as hardy as greater yellowlegs, lesser yellowlegs do not usually appear until early April (February 26, 1996, earliest record) and have moved out of the valley by the middle of May. They are more common during fall migration, returning at the end of June (June 23, 1986, earliest record) and generally leaving by late October. There are a number of November, December, and January records for lesser yellowlegs, but they are much rarer in winter than greater yellowlegs. Late fall and winter records include November 8, 1989, at Old River Reservoir; November 26, 1987, at Stillwater NWR; and November 27, 1987, December 19, 1993, and January 7, 1987, at Carson Lake.

Solitary Sandpiper
(*Tringa solitaria*)
STATUS: Rare but regular migrant.

A handful of solitary sandpipers are found each year from late April (April 10, 1998, earliest spring record) to the middle of May (May 16, 1999, latest spring record). Fall migrants return as early as late June (June 25, 1941), with most records clustering in the second and third weeks of July. The last birds have left by the middle of September (October 1965, latest fall record). Solitary sandpipers are most likely to be found along drains and ditches or small isolated ponds.

Willet
(*Catoptrophorus semipalmatus*)
STATUS: Uncommon breeder and migrant.

Spring migrants usually arrive in the third week of April (April 13, 1987, earliest spring record). Willets are more common breeders and migrants in the Carson Valley–Washoe Valley–north Washoe corridor than in the Lahontan Valley. The differences between these 2 locally distinct migration corridors are not well understood. Fall migrants and breeders are generally gone by the end of July (October 17, 1965, latest fall record).

Spotted Sandpiper
(*Actitis macularia*)
STATUS: Uncommon breeder and migrant.

This species generally arrives in April (February 3, 1968, earliest record) and is gone from the valley by early October (October 18, 1964, latest record). Spotted sandpipers seem to have decreased in the valley, particularly as breeders. Alcorn (1988: 145) reported them as "frequently seen" in summer. The reduced flows in the lower Carson River may have greatly reduced their numbers, as they are common along the nearby lower Truckee River. Neel reported no brood during the 1987–93 drought due to loss of habitat. High count: 150 birds, May 25, 1968, at Stillwater WMA.

Whimbrel
(*Numenius phaeopus*)
STATUS: Rare spring and fall migrant.

Spring whimbrel sightings are concentrated in the last week of April (April 25, 1997, earliest record) and first 3 weeks of May. Fall migration tends to be more dispersed, with fall records occurring between late June and early October. Most records are for Carson Lake, but whimbrels can sometimes be found feeding in flooded agricultural fields. Usually seen as single birds or in small flocks of 3–4 birds; on occasion single birds associate with flocks of long-billed curlews. High count: 8, April 25, 1997, at Carson Lake.

Long-billed Curlew
(*Numenius americanus*)
STATUS: Uncommon summer resident and migrant.

This species generally arrives in early April (February 7, 1966, earliest spring record); migrant numbers peak into early May. The num-

Long-billed Curlew

ber of breeders varies with habitat availability; 30–40 breeding pairs scattered between Stillwater NWR and Carson Lake is the usual number. Curlews form flocks exceeding 100 birds (high count: 240, July 3, 1995) just after the end of the breeding season (early to mid-July). These large flocks do not linger long in the valley, although some may persist into early winter, including the following records: December 16, 1990, and December 19 in 1993, 1994, and 1997, all at Carson Lake. The territorial flight of the male long-billed curlew is one of the most hauntingly beautiful wildlife spectacles of the Lahontan Valley spring.

Marbled Godwit
(*Limosa fedoa*)
STATUS: Common migrant.

Godwits generally arrive in the valley in late March; most leave by early May, although a few late stragglers sometimes persist to early June (June 6, 1991, latest spring record). The earliest returnees show up the first of July, but it is doubtful that many, if any, summer in the valley. Godwits generally depart by the middle of October, although 60 were present on November 10, 1997, at Stillwater NWR. Highly unusual was a bird sighted December 19, 1993. High count: 1000, September 5, 1949, at Stillwater NWR. Marbled godwits are birds of the deeper turbid waters such as are found on the "Closed Area" units of Stillwater NWR.

Ruddy Turnstone
(*Arenaria interpres*)
STATUS: Extremely rare migrant.

During high-water years, ruddy turnstones occur in small numbers in dowitcher flocks. Not expected annually. Spring records are from the first 2 weeks of May (e.g., May 4, 1996, May 7, 1986, and May 13, 1988, all at Carson Lake). Most records from the fall are from late July through late August, with a cluster during the last week of July. High count: 22, July 30, 1987, at Carson Lake.

Black Turnstone
(*Arenaria melanocephala*)
STATUS: Vagrant.

The single record of a bird observed at Carson Lake on April 23, 1998 (DT, RF, JT, FN 52:363), has been accepted by the NBRC. It is one of few interior records for this species.

Red Knot
(*Calidris canutus*)
STATUS: Rare.

Six records, including 4 birds sighted 6 miles north of Fallon on May 14, 1973 (AB 27:801); 2 birds (1 collected) at Nutgrass unit, Stillwater NWR, on May 16, 1950 (Marshall 1951); a sight record of a single bird on October 4, 1997, at Stillwater NWR (RF); up to 10 at Carson Lake on May 8–12, 1999 (GC, MOB); and 1 bird on April 25–29, 2000 (LN, TF). Red knots may go undetected in the vast dowitcher flocks in the Lahontan Valley.

Sanderling
(*Calidris alba*)
STATUS: Rare spring migrant; rare but regular fall migrant.

Spring records range from the middle of April (March 10, 1941, earliest spring record) to the middle of May (May 18, 1942, latest spring record). A diligent search of sandpiper flocks at Carson Lake or Stillwater NWR in August (July 26, 1941, earliest fall record) and September (October 18, 1950, latest fall record) often turns up a sanderling, and flocks of them are found on occasion. Fall migration peaks in September.

Semipalmated Sandpiper
(*Calidris pusilla*)
STATUS: Rare but regular migrant
(more common in fall than in spring).

Semipalmated sandpipers probably visit the Lahontan Valley more frequently than records indicate. The difficulty of gaining close access to large sandpiper flocks on the expansive mudflats may account for the relatively few records. There are 2 spring records: April

30, 1994, at Carson Lake (JJ); and a bird photographed on May 7, 1998, at Big Soda Lake (BS, MI). Fall migrants pass through between the second week of August (August 6, 1998, earliest fall record) and the middle of September (September 18, 1994). Most records are for 1–3 birds, but Lehman recorded 5 birds on August 10, 1988.

Western Sandpiper
(*Calidris mauri*)
STATUS: Common migrant.

Numbers peak between April 15 (March 11, 1988, earliest spring record) and May 1; almost all birds are gone from the valley by May 10 (May 16, 1998, latest spring record). Fall migration is more protracted and begins in mid-July (July 12, 1997, earliest fall record), peaking in late August, with small groups and individuals persisting into October (November 26, 1987, latest fall record). Winter records are extremely rare (December 18, 1993, winter record). During migration, western sandpipers often occur in large mixed flocks with least sandpipers and dunlins; these mixed flocks are often predomi-

Western Sandpiper

nantly westerns. Largest reported mixed flock: 30,000, May 5, 1955, at Stillwater NWR. More recent peaks have been lower (e.g., 19,000 on October 10, 1986, at Carson Lake).

Least Sandpiper
(*Calidris minutilla*)
STATUS: Common migrant.

The occurrence of least sandpipers largely parallels that of western sandpipers. Spring numbers peak between April 15 (March 11, 1988, earliest spring record) and May 1, and almost all are gone from the valley by May 10. Fall migration, beginning in mid-July (July 12, 1997, earliest fall record), is more protracted. Numbers peak in late August, with small groups and individuals persisting into December. Most large sandpiper flocks that pass through the valley are comprised of 10–30 percent least sandpipers. The smaller flocks that occur along the margins of regulating reservoirs or drain ditches are primarily least sandpipers. While westerns prefer the wide-open

Least Sandpiper

mudflats of the primary wetland units, least sandpipers are birds of flooded stubble, where they may be scattered singly or in groups of 10–20. The least sandpiper is one of the hardiest of the shorebirds; a few individuals or small groups linger into winter (e.g., 10 at S-Line Reservoir on January 9, 1997). In mild winters, larger flocks have been reported (e.g. 200+ at S-Line Reservoir on December 15, 2000).

Baird's Sandpiper
(*Calidris bairdii*)
STATUS: Rare to uncommon migrant; more regular in fall than in spring.

Extremely few spring records, including 4 on April 29, 1998, from Stillwater NWR and 1 on May 17, 1992, from Big Soda Lake. Baird's are most often encountered in small numbers in mid to late August, peaking between the 10th and 20th. It is not unusual to encounter this sandpiper in late fall; the latest record is November 26, 1987, at Stillwater NWR. A flock of 17 on August 20, 1998, at East Alkali Lake, Stillwater NWR, is the most ever seen at one time. A reported count of 200 on September 24, 1972, at Stillwater WMA we consider unlikely.

Pectoral Sandpiper
(*Calidris melanotos*)
STATUS: Rare to uncommon migrant, more regular in fall.

There are only a few spring records, mostly from April (March 29, 1988, earliest spring record; May 13, 1988, latest spring record). Fall migration extends from July through October (July 7, 1989, earliest record; December 7, 1993, latest fall record). Pectorals most often occur in small numbers, often with least sandpipers in flooded stubble. High count: 24, September 25, 1996, at Big Soda Lake.

Sharp-tailed Sandpiper
(Calidris acuminata)
STATUS: Vagrant.

A juvenile photographed on August 18, 1996, at Carson Lake (DT, JW) is the earliest record for a juvenile of this species recorded in the lower 48 states. This second record for Nevada is accepted by the NBRC.

Dunlin
(Calidris alpina)
STATUS: Common spring migrant; uncommon fall migrant.

Spring migration usually peaks in late April and early May, but individuals have been seen much earlier (February 26, 1996, March 19, 1951, and March 25, 1994). Most have left the valley by the second week of May (May 16, 1998, latest spring record). Spring peaks include 11,136 birds counted April 21–23, 1989, and 9369 on April 24–30, 1990. Fewer dunlin migrate through the valley in the fall, usually late September–October (August 8, 1988, earliest fall record). Weather permitting, they remain into the early winter. Winter records include a flock of 50 at Carson Lake on December 19, 1993 (LN, WH, AJ); 77 on December 17, 2000, at Carson Lake; 6 on January 9, 1997, at S-Line Reservoir; and 1 specimen from January 14, 1955, at Stillwater NWR (USNM no. 463400).

Curlew Sandpiper
(Calidris ferruginea)
STATUS: Vagrant.

A breeding-plumage curlew sandpiper was photographed at Big Soda Lake on May 7, 1995 (JW) (FN 49:281). It was present with dunlins and constitutes the first Nevada record for this species.

Stilt Sandpiper

(*Calidris himantopus*)

STATUS: Vagrant.

A stilt sandpiper was sighted on July 30, 1939, between Hazen and Mahala (Slipp 1942). An adult bird observed August 11–12, 1988, at both Stillwater NWR and Carson Lake may have been a single individual (PL).

Ruff

(*Philomachus pugnax*)

STATUS: Vagrant.

A female was sighted on April 30, 1994, at Carson Lake (DM, GC, KG); and a male was photographed on April 20, 1997, at Carson Lake (LC) (FN 51:901). These records were the second and third records for Nevada and have been accepted by the NBRC. There is an additional sight record from August 18, 1997, from Carson Lake (LN, WH).

Short-billed Dowitcher

(*Limnodromus griseus*)

STATUS: Rare migrant, although probably overlooked.

The true number of short-billed dowitchers that pass through the valley is unknown. The odd individual (or sometimes a small flock) is betrayed most often by its mellow *tu-tu-tu* call. The short-billed dowitcher is more often encountered in fall than in spring. Spring migrants are found in April (April 26, 1990, and April 6, 2000); fall records cluster from the middle of August (July 3, 1999, earliest fall record) to the middle of September (September 13, 1994, latest fall record).

Long-billed Dowitcher

Long-billed Dowitcher
(*Limnodromus scolopaceus*)
STATUS: Common migrant.

Spring migration begins in the middle of March (February 2, 1995, earliest spring record) and peaks in the last 2 weeks of April. Spring migrants are largely gone by the end of the first week of May (May 14, 1994, latest spring record). Alcorn (1988) reported 4 records from June without dates. These possibly constitute early fall migrants. Fall migrants begin to return in the middle of July (July 9, 1993, earliest fall record). Although numbers generally peak in late August, large numbers may remain into September; 34,000 were counted on September 30, 1968, at Stillwater WMA. Small flocks remain into early winter (November 4, 1986; 134 birds counted on November 11, 1997, and December 19, 1993).

This is the species responsible for the Lahontan Valley wetlands' designation as part of the Western Hemispheric Shorebird Reserve Network in 1988. One-day counts at Carson Lake and Stillwater NWR have reached as high as 100,000 (May 4, 1990). At one point in the 1980s, shorebird scientists speculated that the Lahontan Valley might refuel more than 30 percent of all the migrating long-billed dowitchers in the world.

Common Snipe
(*Gallinago gallinago*)
STATUS: Uncommon summer resident and migrant; uncommon winter resident.

This species is least prevalent during June and July but does breed in wet meadow habitat in the Lahontan Valley and northern Nevada. Snipe are most common in fall. Their distribution in the valley is very localized in semipermanently flooded salt grass or other flooded pasture. Carson Lake, the Canvasback Gun Club, Sheckler Reservoir (when watered), and Harmon Reservoir are regularly visited by this species. High count: 80, August 18, 1997, at Carson Lake.

Wilson's Phalarope
(*Phalaropus tricolor*)
STATUS: Uncommon summer resident; common migrant.

Spring migration begins in late April (February 26, 1996, earliest spring record) and peaks in May. If the water conditions are right, phalaropes will breed in the valley. Stillwater WMA reported 3150 young produced in 1973 (AB 27:899). Platou documented nesting in 1986, but no nesting between 1986 and 1994 (Herron 1986). Nesting resumed in 1994 at Carson Lake with 34 nests; in 1997 50+ nests were recorded (LN). It is believed that restoration of the wetlands through prime water purchases will restore Wilson's phalarope as a regular

Wilson's Phalarope

Lahontan Valley breeder. Fall migration begins in the third week of June as females arrive from more northerly breeding grounds and peaks in early July through the middle of August. Smaller numbers linger into September (October 11, 1961, latest fall record).

High counts: 40,000, July 30, 1953; and 67,000, July 2, 1967, both at Stillwater NWR. More recent high counts include 12,200 on June 25, 1987, at Carson Lake.

Red-necked Phalarope

(*Phalaropus lobatus*)

STATUS: Common migrant.

Spring migrants arrive from late April (April 21, 1989, earliest spring record) to mid-May (May 22, 1960, latest spring record). Fall migrants return in early July (July 7, 1989, earliest fall record). Fall migration peaks in July through the middle of September, with large concentrations at Big Soda Lake. Small numbers remain until early October (October 12, 1949, latest record). High fall count: 6000, July 12, 1968, at Stillwater WMA. Alcorn collected a specimen on June 3, 1991 (NHM no. 28977).

Red Phalarope
(*Phalaropus fulicaria*)
STATUS: Vagrant.

The 3 sight records include a juvenile at the sump at Carson Lake on September 7, 1995 (GC, LN) (AB 50:86). The latter record is accepted by the NBRC and constitutes the first documented northern Nevada record. In addition, 1–2 birds were at Big Soda Lake September 17–18, 1998 (RS); and 1 late bird was at Big Soda Lake on November 28, 1998 (DS, JW).

FAMILY LARIDAE:
JAEGERS, GULLS, AND TERNS

Parasitic Jaeger
(*Stercorarius parasiticus*)
STATUS: Vagrant.

A dark-morph juvenile sighted on Lott Freeway at Carson Lake on August 28, 1999 (DS), appeared either sick or injured and was unable to fly. On the same day another jaeger (unidentified to species) was seen at Carson Lake chasing Caspian terns (DS). In addition, a sick juvenile was picked up at Soda Lake, and later died, on August 29, 2000 (WH). The specimen is at the Stillwater NWR office in Fallon.

Pomarine Jaeger

(*Stercorarius pomarinus*)
STATUS: Vagrant.

One sight record of a dark-phase adult was reported from Doghead Pond, Stillwater NWR, June 25, 1991 (AB 45:1142).

Long-tailed Jaeger

(*Stercorarius longicaudus*)
STATUS: Vagrant.

One record: an adult bird photographed at Carson Lake. The bird arrived July 15 and remained at least through August 22, 1989 (AB 43:1347). An adult long-tailed jaeger seen September 28, 1989, at Stillwater NWR was assumed to be the same bird.

Franklin's Gull

(*Larus pipixcan*)
STATUS: Rare to uncommon spring and fall transient;
irregularly common in summer.

The occurrence of this species appears tied to water conditions in the wetlands. In low-water years small numbers are sighted during spring and fall migrations. Spring migrants generally arrive at the beginning of May (April 17, 1996, and 1999, earliest records) and depart by late August (September 17, 1998, latest record). If water conditions in the wetlands are good during August, migrant Franklin's gulls stage at Stillwater NWR and Carson Lake for several weeks. High count: 238, August 19, 1997, at Stillwater NWR.

Franklin's gulls were first reported nesting in the Lahontan Valley in 1980 and have nested sporadically since that time (AB 35:964). The highest recorded nest total is 50 in 1980. During the 1987–93 drought none were reported nesting, and it was not until the wetlands recovered in the summer of 1995 that this species nested in the valley again.

Franklin's Gull

Bonaparte's Gull
(*Larus philadelphia*)
STATUS: Uncommon spring and fall transient;
rare summer resident.

Bonaparte's gulls may occur in the valley almost anytime between
April (April 4, 1997, earliest record) and November (December 12,
1990, latest record), but they are probably most numerous in early
May and late August. In some years, immature birds are observed in
early July.

Heermann's Gull
(*Larus heermanni*)
STATUS: Vagrant.

Two records: a sight record of a adult flying by Gull Island at Lahontan Reservoir on June 3, 1989 (DM, HJ); and an adult photographed breeding with a California gull on Gull Island May 19–June 3, 1990 (HJ). There are several records from Pyramid Lake (Washoe County) and Virginia Lake (Washoe County).

Ring-billed Gull
(*Larus delawarensis*)
STATUS: Common spring and fall transient;
common summer resident; uncommon winter resident.

Although small numbers overwinter in the Lahontan Valley, ring-billed gulls return in the middle of March in large numbers. Summer birds are joined by fall migrants beginning in August. In the middle of November the number of ring-billed gulls gradually decreases.

Ring-billed gulls only recently colonized this area. In 1946, Alcorn reported none nesting in the valley; but by May 1959 he reported 13 nests at Lahontan Reservoir. Numbers of nesting pairs on the islands of Lahontan Reservoir have increased since the mid-1980s, with 300 nests reported in 1987. Lahontan Reservoir is the only nesting colony in the valley.

The ring-billed gull is common in urban areas, wetlands, irrigated fields, reservoirs, and the dump. During the summer months, ring-billeds are greatly outnumbered by California gulls; outside the nesting season, both species are equally common.

California Gull

California Gull
(*Larus californicus*)
STATUS: Common spring and fall transient;
common summer resident; uncommon winter resident.

Most California gulls arrive in late March or early April, and they remain common to abundant throughout the summer. They nest in stable numbers at Lahontan Reservoir and Anaho Island at Pyramid Lake (e.g., 1300 nests on June 3, 1997). California gulls are opportunistic nesters (e.g., in 1986–87 up to 3600 pairs nested on flooded dunes in the Carson Sink).

California and ring-billed gulls are often seen in newly flooded wetland units and hay fields. As the flood spreads across a dry unit, it pushes insects before it, creating a concentrated line of prey that greatly facilitates gull foraging.

Herring Gull

(*Larus argentatus*)

STATUS: Uncommon spring and fall transient; uncommon winter resident.

Herring gulls occur in the valley from late November through early March (March 28, latest record), although it is unusual to see more than a handful in the flocks of winter gulls. The high count of 17 was observed during spring migration on March 28, 1988. Herring gulls are far more common at nearby Walker and Pyramid Lakes, where congregations of a couple of hundred birds have been reported. In the valley, Lahontan Reservoir, the Fallon dump, and Harmon and S-Line Reservoirs provide the most often used habitat. There is a rare summer record from July 30, 1997, at Stillwater NWR (RF).

Thayer's Gull

(*Larus thayeri*)

STATUS: Vagrant; possible rare winter visitant.

The only record is of an adult and a first-winter bird at S-Line Reservoir on February 26, 1996 (GC, LN); however, this species is a rare but regular winter visitor at Pyramid Lake and Virginia Lake (Washoe County).

Glaucous Gull

(*Larus hyperboreus*)

STATUS: Vagrant.

There are 3 records: November 1, 1988, at Harmon Reservoir; December 6, 1993, at the Fallon dump; and February 21, 1994, at S-Line Reservoir. There are additional winter records from Pyramid Lake and Virginia Lake in Reno.

Sabine's Gull
(*Xema sabini*)
STATUS: Rare fall migrant.

There are a total of 7 records for the Lahontan Valley: 3 individuals seen on August 13–16, 1994 (AB 49:75), 1 on September 17, 1999 (MM), 4 immatures on September 18, 1998 (MM, GS), a single bird on September 22, 1995, and September 26, 1940, all from Big Soda Lake; single birds were also reported on September 28, 1981, at Lahontan Reservoir; and October 7, 1954, on Pintail Bay at Stillwater NWR.

Caspian Tern
(*Sterna caspia*)
STATUS: Uncommon transient; uncommon summer resident.

Spring birds generally arrive in the middle of April (March 21, 1997, earliest record), and fall migrants depart by late September (October 3, 1995, latest record). Caspian terns are opportunistic nesters in the Lahontan Valley, taking advantage of habitat conditions created when high runoff fills the Carson Sink. During the high-water years of the mid-1980s, 475 nests were counted in 1986 and 110 nests were counted in 1987.

Common Tern
(*Sterna hirundo*)
STATUS: Rare spring and fall migrant.

Migrating birds pass through mostly in September, with records falling between September 4 and 24, primarily from Little Soda Lake. During fall migration it is not unusual to see common terns in small flocks. Although only 1 record exists for spring (May 18, 1971, at Stillwater NWR), it is suspected that they have been overlooked at this season.

Forster's Tern

Forster's Tern
(*Sterna forsteri*)
STATUS: Common summer resident.

This species normally appears in the valley in mid-April (April 10, 1968, earliest record) and departs in late September. Numbers have dropped since the 1950s from approximately 500 pairs to about 200 pairs. This decline is probably linked to loss of habitat resulting from the elimination of winter hydropower flows from Lahontan Reservoir. Forster's terns returned to the Lahontan Valley as a major nesting species in 1995 after a 9-year period of low nesting activity; 150–200 pairs were scattered around the primary wetlands in separate colonies of 30–50 pairs. The recent high count of 538 birds was made August 18–19, 1997, at Carson Lake and Stillwater NWR.

Black Tern

Black Tern

(*Chlidonias niger*)

STATUS: Common migrant; uncommon summer resident.

The first migrants arrive in late April, but most arrive in May (April 16, 1992, earliest spring record). Black terns tend to leave the valley soon after the young are fledged; most birds are gone by September 1. After almost disappearing as a breeding species from 1987 to 1993, black terns returned to the valley in higher numbers in 1994, although breeding numbers remain depressed. This species may never attain the breeding numbers of the 1950s (e.g., 170 nests were reported from Pelican Island, Stillwater WMA, on July 18, 1952), when hydropower water releases from Lahontan Dam in winter created extensive permanent, freshwater marshes.

FAMILY ALCIDAE: MURRELETS

Ancient Murrelet
(*Synthliboramphus antiquus*)
STATUS: Vagrant.

A dead bird was found 4 miles west of Fallon April 5, 1970, by C. J. Chamberlain (BMNH no. MBM 1295). There are 2 records from Pyramid Lake just northwest of the Lahontan Valley.

FAMILY COLUMBIDAE: PIGEONS AND DOVES

Rock Dove
(*Columba livia*)
STATUS: Nonnative; common permanent resident.

Rock doves nest on cliffs and rock formations in the ranges immediately surrounding the valley, as well as in various urban and farm buildings in the valley. Winter flocks are often associated with feedlots and dairies, where the birds subsist on waste grain.

Band-tailed Pigeon
(*Columba fasciata*)
STATUS: Irregular transient.

Band-tailed pigeons tend to spread into the valleys and ranges east of their normal Sierra Nevada haunts when pinyon nut crops are good. Records from the Lahontan Valley include 6 birds on May 22, 1983, 6 miles west of Fallon; 1 bird on July 5, 1972, in Fallon; 1 bird on July 12, 1971, along the lower Carson River; and 1 bird on October 17, 1940, 4 miles west of Fallon (MVZ CA no. 80590).

Birds of the Lahontan Valley

White-winged Dove
(*Zenaida asiatica*)
STATUS: Vagrant.

Five records, including 1 specimen. Alcorn (1988) reported identifying a white-winged dove based on a wing saved by a hunter from a bird shot in the Fallon area in September 1946; and Jim Curran collected 1 bird on September 1, 1977 (MVZ CA no. 165063), along Pasture Road south of Fallon. In fall 2000, 4 birds were recorded, including 2 immature birds shot at Carson Lake on September 2, 1 observed 15 miles east of Fallon on October 22 (GT, JT), and 1 at a feeder south of Fallon from November 30 through December 19 (WH).

Mourning Dove
(*Zenaida macroura*)
STATUS: Abundant summer resident;
uncommon winter resident.

Flocks winter locally in the valley. Mourning doves are found throughout the valley where trees are found, but particularly in agricultural and riparian areas.

FAMILY CUCULIDAE: CUCKOOS

Yellow-billed Cuckoo
(*Coccyzus americanus*)
STATUS: Formerly an uncommon summer resident,
now an extremely rare summer visitant.

This species has suffered serious declines in western Nevada due to the loss of riparian habitat. In 1946, Alcorn reported cuckoos every year from late May through August in the Lahontan Valley, including 9 in July 1941. In 1988 they had become rarer, with only 6 records since 1946 (Alcorn 1988). Since 1988 cuckoos have been recorded in the cottonwood floodplain forest at the upper end of Lahontan Reservoir just outside the Lahontan Valley almost annually. This site is now the only known site for cuckoos in northern Nevada. Extensive surveys in the 1970s were unable to locate any cuckoos on the Walker, Carson, or Truckee Rivers. The most recent record for the Lahontan Valley is of a bird photographed on June 24, 1986, at Carson Lake.

Common Barn Owl

FAMILY TYTONIDAE: BARN OWLS

Common Barn Owl
(*Tyto alba*)
STATUS: Common resident,
often overlooked.

Barn owls are widespread in the valley's agricultural area and in out-lying areas where cliffs provide suitable nesting sites. This species will nest in agricultural outbuildings and in haystacks.

FAMILY STRIGIDAE: TYPICAL OWLS

Flammulated Owl
(*Otus flammeolus*)
STATUS: Vagrant, but perhaps more common than
the 2 records suggest.

The 2 specimens include a female found dead 6 miles west-south-west of Fallon on May 9, 1963 (MVZ CA no. 168774); and one from November 1940 (MVZ NV no. 17).

Western Screech Owl
(*Otus kennicottii*)
STATUS: Uncommon resident, often overlooked.

Present in the Carson River corridor and throughout the valley, par-ticularly where large cottonwoods provide suitable habitat. Screech owls occupy nesting boxes from December through April.

Western Screech Owl

Great Horned Owl

(*Bubo virginianus*)

STATUS: Common resident.

The most common owl species in the Lahontan Valley, great horned owls begin nesting as early as late January, often in abandoned nests built by red-tailed hawks, Swainson's hawks, common ravens, or black-billed magpies.

Northern Pygmy-Owl

(*Glaucidium gnoma*)

STATUS: Vagrant, perhaps more regular.

One sight record from Fallon on December 15, 2000 (BC, BEC). This species probably occurs more regularly during the winter months.

Burrowing Owl

Burrowing Owl

(*Athene cunicularia*)

STATUS: Uncommon summer resident; rare winter resident?

Burrowing owls usually occur in the valley from March (March 6, 1995, earliest record) through October. Alcorn (1988:194) reported that the number of birds has diminished since 1946, "possibly by as much as 50 percent in recent years." He also noted having observed this species in all months. These owls nest in small numbers in the Lahontan Valley, including at Carson Lake and the Indian Lakes area. They are most easily observed along several of the drains at Carson Lake, where they occupy burrows in the drain banks.

Long-eared Owl
(*Asio otus*)

STATUS: Rare resident, but more common as
a migrant and winter resident than in summer.

Nesting in the valley was confirmed on March 31, 1936, and March 27, 1942 (Alcorn 1988). Long-eared owls occur sporadically along the river corridor. They prefer mature black willow thickets, a habitat largely replaced by Russian olive along the riparian corridor below Lahontan Dam. Long-eared owls are more consistently encountered above Lahontan Reservoir.

Short-eared Owl
(*Asio flammeus*)

STATUS: Uncommon winter resident in most years;
rare, but regular, summer resident.

Short-eared owls are most common in the valley from November to February. Nests were located in April 1993 at South Nutgrass, Stillwater NWR, on June 23, 1970, no location; and on June 29, 1993, and July 3, 1995, at Carson Lake. The number of short-eared owls present at any given time is probably related to conditions in the wetlands and may actually increase as vole populations build up during drought periods.

Northern Saw-whet Owl
(*Aegolius acadicus*)

STATUS: Rare winter visitor, but possibly overlooked.

There are 4 records for this species: December 1935 and 1937, individuals found dead 4 miles west of Fallon (Alcorn 1940); January 8, 1960, 1 mile north of Fallon (MVZ CA no. 168775); and January 15, 1966.

FAMILY CAPRIMULGIDAE: GOATSUCKERS (NIGHTJARS)

Common Nighthawk
(*Chordeiles minor*)
STATUS: Common summer resident and migrant.

Common nighthawks arrive in the valley in late May (April 19, 1998, earliest record) and are gone by early September (September 21, 1998, latest record). Alcorn (1988) confirmed that nighthawks nest in the Lahontan Valley. In late August large flocks are often found in and around Fallon.

Common Poorwill
(*Phalaenoptilus nuttallii*)
STATUS: Uncommon migrant and summer resident.

Poorwills arrive in early May (April 17, 1987, earliest record) and are gone by early October. They occur in the sandy uplands along the Carson River from Lahontan Dam to the Carson River Diversion Dam, and in the foothills of the surrounding mountains in areas dominated by sagebrush. They are not found in agricultural and wetland habitats.

FAMILY APODIDAE: SWIFTS

Vaux's Swift
(*Chaetura vauxi*)
STATUS: Rare but regular migrant.

Most common in May (e.g., 2 birds on May 4, 1960; 2 birds on May 8, 1961; 1 bird on May 13, 1995; 1 bird on May 14, 1970; 4 birds on May 16, 1998; 5 birds on May 20, 1959). There is 1 fall record of a single bird from September 7, 1985.

FAMILY TROCHILIDAE: HUMMINGBIRDS

Black-chinned Hummingbird
(*Archilochus alexandri*)
STATUS: Uncommon migrant and summer resident.

Arriving in late April (April 19, 1998, earliest record), black-chinned hummingbirds are probably gone by the end of September (September 27, 1992, latest record). Unfortunately, hummingbirds of any species are not very common in the Lahontan Valley at any time of the year, probably due to the loss of riparian habitat, particularly the replacement of the understory with exotic vegetation along the lower Carson River. Nesting has not been documented in the valley, although a pair was seen consistently at a feeder during the summer of 1997 (RF). Nesting has been documented in nearby Fernley.

Anna's Hummingbird
(*Calypte anna*)
STATUS: Vagrant; possible rare migrant.

One male was observed in Fallon on April 26–29, 1997 (RF).

Costa's Hummingbird
(*Calypte costae*)
STATUS: Vagrant.

One male was observed at a residence in Fallon, May 27–June 5, 1998 (RF). Although in Nevada Costa's hummingbirds are considered to be a species of the southern desert, there are a handful of northern records, including the regular occurrence of a pair near Mullin Pass on the west side of Pyramid Lake (Washoe County). The status of this species needs investigation.

Calliope Hummingbird

(*Stellula calliope*)

STATUS: Uncommon migrant; rare summer resident?

Calliope hummingbirds probably arrive in early May and depart by early to mid-July. This tiny hummingbird inhabits the creeks and springs of Nevada's numerous mountain ranges. In the Lahontan Valley, this species is most often seen in migration, although it has nested here. Alcorn (1988) reported 1–2 present in Fallon in the summer of 1970 and a nest 4 miles west of Fallon in 1968 and 1969. At this same location an adult was seen feeding 2 young on the ground on June 20, 1977.

Broad-tailed Hummingbird

(*Selasphorus platycercus*)

STATUS: Rare to uncommon migrant.

Present as a migrant along the lower Carson River and in Fallon during May and August (August 16, 1997, earliest record) and September.

Rufous Hummingbird

(*Selasphorus rufus*)

STATUS: Uncommon fall migrant; probable spring migrant.

This species has been detected only as a fall migrant beginning in mid-July through September (September 28, 1996, latest record), but may also occur as a spring migrant.

FAMILY ALCEDINIDAE: KINGFISHERS

Belted Kingfisher
(*Ceryle alcyon*)
STATUS: Uncommon year-round resident.

Most commonly observed patrolling the water-delivery canals during the off-season when small fish get stranded in isolated pools. Other consistent haunts include S-Line and Harmon Reservoirs around the outlet structures and the Carson River.

FAMILY PICIDAE: WOODPECKERS

Lewis' Woodpecker
(*Melanerpes lewis*)
STATUS: Rare but regular spring and fall transient.

Migrants usually pass through the valley singly during May and in September (August 27, 1940, earliest fall record) and October (October 29, 1998, latest fall record). This species breeds in the nearby Carson Range in western Nevada.

Acorn Woodpecker
(*Melanerpes formicivorus*)
STATUS: Vagrant.

One sight record from May 8, 1988, along Pasture Road just west of the Greenhead Club (LN, MOB). There is a breeding population near Susanville, California.

Williamson's Sapsucker
(*Sphyrapicus thyroideus*)
STATUS: Vagrant.

There is a sight record of a female from September 16, 1994, at the Carson River Diversion Dam (LN).

Red-naped Sapsucker
(*Sphyrapicus nuchalis*)
STATUS: Rare migrant.

Most records are from April and between mid-September (September 19, 1995, early record) and mid-October (October 20, 1951, latest record) along the Carson River corridor, particularly Timber Lakes. Alcorn collected 2 specimens: April 23, 1941 (NHM no. 28980); and April 17, 1942 (NHM no. 28981). An uncommon to common summer resident in most northwestern Nevada mountain ranges, the red-naped sapsucker can be found associated with aspen, mountain mahogany, or willow riparian habitats. Its closest summer haunts to the Lahontan Valley are riparian areas in the Stillwater Range. Birds found in the valley are in passage and do not typically stay for any length of time.

Red-breasted Sapsucker
(*Sphyrapicus ruber*)
STATUS: Rare fall migrant.

Although this species is a summer resident as far east as the Wassuk Range (60 miles due south of the Lahontan Valley), it is more closely associated with the Sierra Nevada. An occasional migrant strays east of its normal haunts into the Carson River corridor through the Lahontan Valley. Records include September 19, 1988, at Timber Lakes; and November 27, 1987 (PL).

Downy Woodpecker

Downy Woodpecker

(*Picoides pubescens*)

STATUS: Uncommon year-round resident and transient.

This species was formerly more common along the lower Carson River (Alcorn 1988). Downy woodpeckers are found in riparian forest along the Carson River, and less commonly in other wooded areas around farmsteads in the valley.

Hairy Woodpecker

(*Picoides villosus*)

STATUS: Uncommon resident and transient.

Hairy woodpeckers occur commonly along the river corridor and wooded farmsteads just about any time of year but are most abundant in the valley during migration.

Northern Flicker

(*Colaptes auratus*)

STATUS: Common year-round resident.

Northern flickers occur commonly throughout the valley and form impressive concentrations from September to late December when Russian olive crops are heavy. They begin dispersing in early January.

FAMILY TYRANNIDAE:
TYRANT FLYCATCHERS

Olive-sided Flycatcher
(*Contopus borealis*)
STATUS: Uncommon migrant.

Spring migrants are seen in the valley during May, especially late May (April 30, 1960, earliest spring record; June 2, 1960, latest spring record). Fall migration occurs from late August (August 22, 1998, earliest fall record) through September (September 23, 1998, latest fall record). Most likely to be found in the Carson River corridor, particularly around the Carson River Diversion Dam and Timber Lakes.

Western Wood-Pewee
(*Contopus sordidulus*)
STATUS: Common migrant; possible breeder?

Spring migrants transit the valley in May (April 19, 1997, earliest spring record) and early June (June 21, 1995, latest spring record). Fall migrants pass through from August (August 4, 1994, earliest fall record) through early October (October 6, 1994, latest fall record) and commonly occur along the river corridor. Wood-pewees nest in the river corridor above Lahontan Reservoir, but the riparian corridor below Lahontan Dam may provide only marginal breeding habitat.

Willow Flycatcher
(*Empidonax traillii*)
STATUS: Uncommon migrant.

This species probably occurs in small numbers as a spring migrant during the second half of May (May 4, 1997, earliest spring record) and early June and as a fall migrant during August (August 11, 1998, earliest fall record) and early September. A specimen was collected by Alcorn on August 26, 1940 (MVZ CA no. 79409).

Hammond's Flycatcher
(*Empidonax hammondii*)
STATUS: Uncommon to rare migrant.

The occurrence of this species in the valley is not well understood, although recent fall banding indicates that perhaps it is more common than previously assumed. Linsdale reported examining a specimen (USNM no. 298439) collected 4 miles west of Fallon on May 13, 1925 (Linsdale 1936). There is another specimen from May 13, 1960 (MVZ NV no. 1293). Recent banding efforts at Timber Lakes yielded 4 birds: September 4, 1998, September 18, 1998, September 5, 2000, and September 13, 2000 (DW). In addition, a sight record was reported from S-Line Reservoir on May 15, 1999 (DS, JE).

Gray Flycatcher
(*Empidonax wrightii*)
STATUS: Common migrant.

Usually the earliest and most common empidonax migrant, gray flycatchers pass through from late April through the middle of May, and again from August (July 30, 1997, earliest fall record) through September (September 17, 1999, latest fall record). The gray flycatcher is a fairly common breeder in the regions all around the Lahontan Valley.

Dusky Flycatcher
(*Empidonax oberholseri*)
STATUS: Uncommon migrant.

Dusky flycatchers pass through the valley in the second half of May (May 2, 1942, earliest spring record; May 30, 1942, latest spring record) and again in mid-August and early September. They are common breeders in the mountains of western Nevada. Along with gray flycatchers, dusky flycatchers probably make up the bulk of the empidonax flycatchers that pass through the Lahontan Valley on migration. Alcorn collected a specimen on August 11, 1941 (NHM no. 28986).

Western-type Flycatcher
(*Empidonax occidentalis / difficilis*)
STATUS: Uncommon migrant.

Expected as a migrant in mid to late May and again after the middle of August through September (September 28, 1997, latest record). Observers have just started sorting out "western-type" flycatchers in the valley to determine whether they are cordilleran flycatchers or Pacific-slope flycatchers. The banders at Timber Lakes report banding a number of birds that they were unable to definitely identify to species, but of the birds they have been able to identify to species, they report only Pacific-sloped flycatchers, including single birds on August 31, 2000, and September 7, 2000. They have yet to positively identify any cordilleran flycatchers. As passage birds, "western" flycatchers occur most commonly along the Carson River corridor in the Lahontan Valley (e.g., in the vicinity of the Carson River Diversion Dam and Timber Lake).

Black Phoebe
(*Sayornis nigricans*)
STATUS: Rare, irregular visitant.

An uncommon but regular summer resident in Mason Valley to the south, the actual status of the black phoebe in the Lahontan Valley is hard to ascertain. They occur in every season with little apparent pattern. Although the timing of their appearance in the area is unpredictable, they occur with some regularity along the Carson River immediately below Lahontan Dam. Nesting has not been confirmed.

Say's Phoebe

Say's Phoebe

(*Sayornis saya*)

STATUS: Uncommon summer resident; rare winter resident.

One of the earliest passerine migrants, Say's phoebes generally arrive the first week in March. A few hardy birds winter in the valley, and 1 or 2 have been tallied on 8 of 13 Fallon CBCs between 1985 and 1997. Say's phoebes nest in cracks and crevices in the tufa stacks and cliffs in the hills surrounding the valley. In the valley itself, they nest in barns and hangars, and under eaves.

Vermilion Flycatcher
(*Pyrocephalus rubinus*)
STATUS: Vagrant.

Four records: 2 birds (1 taken) 6 miles southeast of Fallon in October 1948; 1 male taken by Alcorn on December 8, 1957 (NHM no. 35325); and 1 male recorded from December 9, 1996. The latter remained in Fallon until April 5, 1997 (photo and documentation accepted by the NBRC).

Ash-throated Flycatcher
(*Myiarchus cinerascens*)
STATUS: Rare local summer resident (formerly more common).

Arrives in early May (May 4, 1997, earliest record) and departs late August (August 26, 1994, latest record). A small population is resident along the Carson River below Sagouspi Dam from early May through August. They have become increasingly scarce in the last 30 years (Alcorn 1988). Although habitat modification has undoubtedly caused this perceived decline, it is difficult to know whether to blame the loss of buffaloberry stands in the valley or the invasion of Russian olive into the cottonwood riparian areas along the river. Ash-throated flycatchers prefer a fully developed mid-story shrub-tree layer for foraging and cavity-bearing overmature cottonwood trees for nesting, a combination that is no longer prevalent in the valley.

Western Kingbird
(*Tyrannus verticalis*)
STATUS: Common summer resident.

Western kingbirds arrive in the third week of April (March 14, 1996, earliest record) and leave by early September (October 6, 1994, latest record). They are common in the valley's agricultural areas and are often seen along roadsides.

Western Kingbird

Eastern Kingbird

(*Tyrannus tyrannus*)

STATUS: Rare spring and fall migrant.

Most records of this species are from June, including June 4, 1970, at Stillwater Marsh; June 4, 1998, at Timber Lakes; June 19, 1979; and June 30, 1968. Additional records include July 29, 1977, at Big Soda Lake; and August 24, 1994, along Pioneer Way near the Carson River.

Loggerhead Shrike
(*Lanius ludovicianus*)
STATUS: Common summer resident;
uncommon winter resident.

The Lahontan Valley, and Stillwater NWR in particular, is a stronghold for loggerhead shrikes, supporting high nesting densities. Alcorn (1988) reported that loggerhead shrikes were more common prior to 1960.

Loggerhead Shrike

Northern Shrike

(*Lanius excubitor*)

STATUS: Rare but regular winter visitant.

This species can be seen in the valley from late November (November 14, 1998, earliest record) through late February (February 23, 1995, latest record). Adults and immatures occur with equal frequency. Northern shrikes are regularly seen in the area south of the Fallon Naval Air Station.

FAMILY VIREONIDAE: VIREOS

Plumbeous Vireo

(*Vireo plumbeus*)

STATUS: Uncommon migrant.

Spring migrants arrive in late April and May (April 11, 1997, earliest spring record; May 18, 1995, latest spring record). Fall migrants are present from August (August 1, 1992, earliest fall record) to early October (October 5, 1989, latest fall record). They are most common along the Carson River corridor.

Cassin's Vireo

(*Vireo cassini*)

STATUS: Uncommon (?) migrant.

The first definite record was a bird banded at Timber Lakes on September 2, 1998 (DW, GA). Cassin's vireo breeds in the Carson Range in western Nevada. The status of this species needs to be better documented.

Warbling Vireo
(*Vireo gilvus*)
STATUS: Common migrant.

Warbling vireos occur in the valley during late April and May (June 1, 1995, latest spring record) and again from August (August 1, 1992, and 1997 earliest fall records) through late September (September 29, 1999, latest fall record). Migrants are common along the Carson River corridor. Alcorn (1988) reported specimens of the *leucopolius* race taken on May 21, 1942, and the *swainsonii* race on September 14, 1942.

FAMILY CORVIDAE:
JAYS, MAGPIES, AND CROWS

Blue Jay
(*Cyanocitta cristata*)
STATUS: Vagrant.

Alcorn (1988) reported 2 seen (1 collected) 2.4 miles west of Fallon on December 14, 1976. We were unable to locate the specimen. In addition, there is a sight record of a wintering blue jay in Fernley during the 1987–88 winter.

Western Scrub-Jay
(*Aphelocoma californica*)
STATUS: Rare, irregular winter resident.

An occasional scrub-jay will make its way down to the valley during the harshest winters. Records are from late September through early May (May 10, 1991, latest record). Nearest nesting occurs in the Stillwater Range.

Pinyon Jay
(*Gymnorhinus cyanocephalus*)

STATUS: Vagrant; possibly irregular visitor.

Three sight records: 2 birds were seen on July 12, 1996, flying over Fallon (RF), 3 were observed on September 4, 2000 (RF), and 4 were observed on September 6, 2000, at Timber Lakes (DW). Pinyon jays breed in the nearby Stillwater Range.

Clark's Nutcracker
(*Nucifraga columbiana*)

STATUS: Probable rare transient.

There are 2 sight records: from the summer of 1970, and March 16, 1973. Clark's nutcrackers occur quite uncommonly in the Stillwater Range and commonly in the Pine Nut Range (Lyon County) and Desatoya Mountains (Churchill and Lander Counties).

Black-billed Magpie
(*Pica pica*)

STATUS: Common year-round resident; possibly more common in summer than in winter.

In recent years, numbers in the Lahontan Valley have been surprisingly stable, with CBC numbers ranging from 421 to 632 between 1985 and 1994.

American Crow
(*Corvus brachyrhynchos*)

STATUS: Uncommon year-round resident.

Crows may not have occurred in the valley prior to cultivated agriculture, and the closing of many smaller dumps and dead animal pits has significantly reduced their numbers (CBC numbers usually 2–8 birds between 1985 and 1994). Peak counts in 1987 (409 birds) and 1990 (139 birds) may have included misidentified common ravens.

Common Raven
(*Corvus corax*)
STATUS: Common year-round resident.

Raven numbers vary greatly. Some of this variation may be attributed to the "control" of ravens, but it might also be the result of natural cycles. CBC numbers range from 47 in 1985 to a high of 447 in 1990, followed by a decline to 26 in 1991.

Common Raven

FAMILY ALAUDIDAE: LARKS

Horned Lark
(*Eremophila alpestris*)
STATUS: Abundant winter resident; common summer resident.

Christmas bird count totals fluctuate from 1926 birds in 1991 to 11,100 in 1994, with an average in the 1985–94 period of 5447. Flocks with more than 500 birds are not uncommon. Horned larks favor barren fields and salt-grass flats in winter.

Horned Lark

FAMILY HIRUNDINIDAE: SWALLOWS

Tree Swallow
(*Tachycineta bicolor*)
STATUS: Common migrant; rare breeder.

One of the earliest passerine migrants to reach the valley; spring migration peaks in late March and April, although the first birds begin appearing in early March (February 19, 1994, earliest record). Fall migration peaks from August through the middle of September (November 26, 1987, latest record). Twenty immature birds were reported on July 1, 1996 (RF). Very unusual were the 3 birds reported on December 20, 1987. High counts include approximately 6000 on March 28, 1988, and an estimated 20,000 on April 4 at Carson Lake. Tree swallows nest along the Carson River upstream of Lahontan Reservoir.

Violet-green Swallow
(*Tachycineta thalassina*)
STATUS: Uncommon migrant.

Spring migrants pass through the valley from the end of March through early May, and in August (July 23, 1992, earliest fall record). They are not known to nest in the valley.

Northern Rough-winged Swallow
(*Stelgidopteryx serripennis*)
STATUS: Common migrant; rare summer resident.

This species begins to arrive in early April (March 4, 2000, earliest spring record), and spring migration peaks from mid-April through mid-May. A few isolated pairs nest along the river corridor and along certain major irrigation delivery canals and drains, particularly those with steep, consolidated banks. The same nesting sites are used year-to-year. Most rough-winged swallows have departed by the end of August.

Bank Swallow

Bank Swallow

(*Riparia riparia*)

STATUS: Common migrant; uncommon summer resident.

Bank swallows arrive in the valley in mid-April (March 29, 1988, and 1997 earliest records), usually mixed in with flocks of cliff, rough-winged, and violet-green swallows. They nest in colonies of up to 200 pairs in certain suitable cut banks along the perimeter of La-hontan Reservoir, down the Carson River below Lahontan Dam, and along certain deep irrigation canals and drains. In late July and early August, bank swallows, supplemented by smaller numbers of cliff swallows, congregate in large numbers at certain locales around the valley. The Stewart Road power lines west of Harmon Reservoir consistently serve as one of these staging sites (e.g., 2400 birds on July 19, 1996). All are gone by early September.

Cliff Swallow

(*Hirundo pyrrhonota*)

STATUS: Common migrant and summer resident.

Like bank swallows, cliff swallows arrive in early April (March 6, 1996, earliest record) and depart by late August (August 26, 1962, latest record). They commonly nest under bridges in the Lahontan Valley.

Barn Swallow

(*Hirundo rustica*)

STATUS: Common migrant; common breeder.

Migrants show up in the valley in mid-March (March 7, no year, earliest record). In early April, local breeders arrive and commence breeding around human structures. After raising a couple of broods, barn swallows flock in late summer much like other swallow species, but persist in the valley into late October and early November. High count: 3000, September 13, 1995, at Carson Lake.

FAMILY PARIDAE: TITMICE

Mountain Chickadee

(*Parus gambeli*)

STATUS: Uncommon to rare, but regular,
migrant and winter visitant.

Usually found in very small numbers every winter along the Carson River corridor, usually between the middle of September (September 8, 1987, earliest record) and early April (April 16, 1945, latest record).

FAMILY AEGITHALIDAE: BUSHTITS

Bushtit
(Psaltriparus minimus)
STATUS: Irregular winter visitant; possible rare breeder.

Most often recorded in late November through mid-May (June 1, 1996, latest spring record). Probably more common formerly, with decline due to loss of buffaloberry and riparian vegetation. Alcorn (1988) reported that they were not abundant in 1946 and had diminished by 1988. It is possible that bushtits are rare breeders in the valley. A flock of 6 was seen on August 8, 1998, along Wildes Road east of Fallon (WH). In addition, a flock of 6 birds was banded at Timber Lakes on August 24, 1998.

FAMILY SITTIDAE: NUTHATCHES

Red-breasted Nuthatch
(Sitta canadensis)
STATUS: Rare but regular migrant.

This species occurs in migration and in winter along the Carson River corridor from August (August 1, 1992, earliest record) through late April (May 7, 1988, latest record).

White-breasted Nuthatch
(Sitta carolinensis)
STATUS: Rare migrant.

White-breasted nuthatches migrate through the valley in February and March (April 12, 1997, latest record), and again in September (September 4, 2000, earliest fall record). Less commonly encountered along the river corridor than the red-breasted nuthatch.

FAMILY CERTHIIDAE: CREEPERS

Brown Creeper
(*Certhia americana*)
STATUS: Rare migrant and winter resident.

Most likely encountered between September and early May along the river corridor or in stands of mature cottonwoods.

FAMILY TROGLODYTIDAE: WRENS

Rock Wren
(*Salpinctes obsoletus*)
STATUS: Uncommon year-round resident, rare in winter.

Rock wrens occur on the margins of the valley where minor faulting and volcanic activity have exposed broken cliffs and strewn the fans with basalt boulders. One or 2 pairs usually can also be found in the center of the valley on the skirts of Rattlesnake Mountain.

Canyon Wren
(*Catherpes mexicanus*)
STATUS: Rare resident.

Canyon wrens are rare close to the valley floor. Only 2 sites where this species occurs are known: the low cliffs just southeast of Stillwater Point Reservoir and on Upsal Hogback north of Fallon. They occur more regularly at cliff sites in the surrounding mountains.

Bewick's Wren

(*Thryomanes bewickii*)

STATUS: Common year-round resident.

Common in thick, woody tangles along the river corridor and other shrubby thickets; also in areas with Russian olive and tamarisk (e.g., eastern edge of Harmon Reservoir).

House Wren

(*Troglodytes aedon*)

STATUS: Uncommon to common summer resident and migrant.

House wrens arrive in late April (April 21, 1993, earliest spring record) and depart by early October (October 12, 1994, latest fall record) and thrive in mature riparian forest. Unlike Bewick's wrens, house wrens prefer areas with dead and downed cottonwoods and fewer dense shrubs. As cavity nesters, house wrens are also somewhat limited by the distribution of snags throughout the preferred habitat.

Winter Wren

(*Troglodytes troglodytes*)

STATUS: Vagrant; possibly a rare but regular migrant.

The 4 records include a specimen collected by Alcorn on April 5, 1939, 4 miles west of Fallon (MVZ CA no. 76193); 1 banded at Timber Lakes on September 23, 1998 (DW, GA); a sight record of 2 birds at the Carson River Diversion Dam on November 8, 1998; and a second sight record from the same location on November 26, 1987 (PL).

Marsh Wren

(*Cistothorus palustris*)

STATUS: Common summer resident; fairly common winter resident.

Almost no tule or cattail patch is too small to hold at least 1 marsh wren. Abundant at Carson Lake and Stillwater Marsh from March to November; less numerous during the winter months. Also found in vegetated drain ditches and seepage wetlands around regulating reservoirs.

Marsh Wren

FAMILY CINCLIDAE: DIPPERS

American Dipper
(*Cinclus mexicanus*)
STATUS: Vagrant.

Two sight records: November 26, 1939, at Lewis' Drop; and "mid-November" 1973, 3 miles west of Fallon (Alcorn 1988).

FAMILY REGULIDAE: KINGLETS

Golden-crowned Kinglet
(*Regulus satrapa*)
STATUS: Irregular migrant and winter visitant.

Most records are from the Carson River corridor from October through early December, but some golden-crowned kinglets occur in the valley through March (March 28, 1988, latest record).

Ruby-crowned Kinglet
(*Regulus calendula*)
STATUS: Common migrant; rare but regular winter resident.

Beginning in September (September 11, 1997, earliest fall record), ruby-crowned kinglets are very common; their numbers dwindle as winter progresses until only a few remain. Recorded on 8 of 13 Fallon CBCs between 1985 and 1997. Spring migrants begin to appear in March and have departed by May (May 11, 1995, latest spring record).

FAMILY SYLVIIDAE: GNATCATCHERS

Blue-gray Gnatcatcher
(*Polioptila caerulea*)
STATUS: Rare migrant and summer resident.

Migrants usually pass through in small numbers in late April (April 27, 1942, earliest record) and early May, and again in August–September (September 23, 1994, latest record). Perhaps more common formerly, this species is now a rare breeder in the valley in willows and Russian olive at Timber Lakes. In the lowlands of Nevada, blue-gray gnatcatchers occur in areas with buffaloberry and black willow, and can be regularly encountered in Mason Valley, along the Carson River above Lahontan Reservoir, and along the Truckee River. In the Lahontan Valley, gnatcatchers are most often seen in the Carson River corridor below Sagouspi Dam.

FAMILY TURDIDAE:
SOLITAIRES, THRUSHES, AND ALLIES

Western Bluebird
(*Sialia mexicana*)
STATUS: Uncommon migrant; rare winter resident;
very rare summer resident below Lahontan Dam.

The western bluebird is much more common along the Carson River above Lahontan Reservoir than it is below the dam. Although Alcorn (1988) referred to western bluebirds as resident, he did not specifically refer to nesting records. A nest confirmed on July 5, 1995, along the Carson River immediately below Timber Lake may be the first nesting record documented for the Carson River below Lahontan Dam since the mid-1960s, and coincides with an increase in winter observations since 1992.

Mountain Bluebird
(*Sialia currucoides*)
STATUS: Uncommon winter visitor.

This species winters locally in the valley, particularly along the southern margins of Carson Lake from November (October 17, 1994, earliest record) through January. Migrant waves pass through from early February through early March (March 17, 1987, latest record). Alcorn (1988:275–76) reported "a sharp decline in the population" of mountain bluebirds during the 1970s and 1980s.

Townsend's Solitaire
(*Myadestes townsendi*)
STATUS: Rare migrant and winter resident.

Townsend's solitaires occur almost annually in the valley from September (September 14, 1998, earliest fall record) through early May, with most records from late January through the middle of February. Alcorn (1988) reported an August record with no date.

Swainson's Thrush
(*Catharus ustulatus*)
STATUS: Uncommon migrant.

Swainson's thrush is probably an overlooked migrant along the Carson River corridor during mid to late May (May 30, 1945, latest spring record) and again in late August through September (September 18, 1998, latest fall record).

Hermit Thrush
(*Catharus guttatus*)

STATUS: Uncommon migrant and rare winter resident.

Spring migrants occur from late April through May. Fall migration is poorly known but assumed to begin in September (September 28, 1997, earliest record) and to continue into October. Hermit thrushes have not been reported in the valley during the winter months.

American Robin
(*Turdus migratorius*)

STATUS: Very common summer resident; uncommon winter resident.

Winter numbers vary, perhaps depending on the availability of Russian olive fruit. During the 1985–94 period, Fallon CBC numbers ranged from 12 to 79 birds, except for 368 in 1989 and 240 in 1994. Robins are normally found in town, around homesteads, and along the Carson River corridor.

Varied Thrush
(*Ixoreus naevius*)

STATUS: Irregular fall and winter visitor.

This unexpected visitor may be encountered from October (October 5, 1989, earliest fall record) through the end of March (April 11, 1997, latest spring record), but most records are from October and November.

FAMILY MIMIDAE: MOCKINGBIRD, THRASHERS, AND ALLIES

Northern Mockingbird
(*Mimus polyglottos*)
STATUS: Very uncommon summer resident;
very rare winter visitor.

Northern mockingbirds breed in greasewood communities and in cultivated areas in the valley. This species is also found in greasewood during the winter months.

Sage Thrasher
(*Oreoscoptes montanus*)
STATUS: Uncommon summer resident and migrant;
probable rare winter resident.

Sage thrashers arrive in late March and begin departing in late August; most are gone by late October. Although there are few winter records to confirm it, this species probably overwinters during mild winters. Sage thrashers appear to be limited to sites in the valley with shrub-covered dunes, such as the areas around Indian Lakes and Massie and Mahala Sloughs and the sagebrush zone south of the Carson River Diversion Dam. They are found more commonly in the upland areas surrounding the valley.

Sage Thrasher

FAMILY STURNIDAE: STARLINGS

European Starling
(*Sturnus vulgaris*)
STATUS: Nonnative; abundant year-round resident.

This is perhaps the most common species around agricultural areas, particularly feedlots. Fallon CBC numbers range from 5737 (1985) to 37,956 (1997). Feedlots and dairies periodically implement starling control to reduce their feed losses.

The first sighting for the Lahontan Valley was of a flock of 22 on February 2, 1947, in the Soda Lake district (Marshall and Alcorn 1952). Alcorn reported that their numbers "increased each year until the winter of 1963–64, [then] . . . declined for several years. The decline may have been due to control operations carried out by the U.S. Fish and Wildlife Service in Lovelock, Fallon and southern Nevada areas" (Alcorn 1988).

FAMILY MOTACILLIDAE: WAGTAILS AND PIPITS

American Pipit
(*Anthus rubescens*)
STATUS: Common winter resident.

American pipits, like horned larks, are winter residents from October (October 26, 1995, earliest record) to late April (May 11, 1971, latest record) in the barren salt-grass flats surrounding the wetlands. Recorded annually on the Fallon CBC from 1985 to 1994 with totals ranging from 2 (1992) to 524 (1997).

FAMILY BOMBYCILLIDAE: WAXWINGS

Bohemian Waxwing

(*Bombycilla garrulus*)

STATUS: Rare and irregular winter visitor.

Bohemian waxwings occur in the Lahontan Valley between November and January, particularly around gardens in town.

Cedar Waxwing

(*Bombycilla cedrorum*)

STATUS: Irregular migrant and winter visitor.

This species generally occurs from October (September 4, 1998, earliest record) through May (June 16, 1995, latest record) in flocks of up to 100 birds. The achenes of the Chinese elm, usually produced in early April before leaf-out, attract waxwing flocks.

FAMILY PTILOGONATIDAE: SILKY FLYCATCHERS

Phainopepla

(*Phainopepla nitens*)

STATUS: Vagrant.

The 3 valley records, 2 of them from 1943, include a sight record from August and a reported specimen from October 11 (Alcorn 1988). We were unable to locate this reported specimen. The most recent is a sight record of a male on May 14-15, 1999, at S-Line Reservoir (LN, MOB). There are at least 4 records of this species from the Mono Lake basin in September and October (Gaines 1988) as well as an unspecified record from nearby Storey County (Ryser 1985).

FAMILY PARULIDAE: WOOD-WARBLERS

Orange-crowned Warbler

(*Vermivora celata*)

STATUS: Common migrant; rare, but regular, winter resident.

Generally encountered in spring migration beginning in early March (March 10, 1996, earliest spring record) through the middle of May (May 18, 1995, latest spring record). Orange-crowned warblers are often a significant component of mixed warbler flocks found along the river corridor in late August through late September. Recorded on 6 of the 13 Fallon CBCs from 1985 to 1997. During the winter months this warbler is most likely to be found in ornamental vegetation around dwellings or in willow thickets along irrigation drains and ditches.

Nashville Warbler

(*Vermivora ruficapilla*)

STATUS: Rare migrant, most likely to be seen during fall migration.

This species moves through the valley in small numbers during early and mid-May and again from August to early September along the Carson River corridor.

Virginia's Warbler

(*Vermivora virginiae*)

STATUS: Vagrant; probably a rare but regular migrant.

One individual was banded at Timber Lakes on September 6, 1999 (RN, DW).

Northern Parula

(*Parula americana*)

STATUS: Vagrant.

There is 1 sight record: a male singing on June 11, 1997, at the edge of the Carson Lake pasture; seen by 2 experienced observers (KB, CH).

Yellow Warbler

(*Dendroica petechia*)

STATUS: Common migrant; uncommon summer resident.

Yellow warblers generally arrive in the valley the first week of May and are gone by mid-September (September 24, 1994, and 1998 latest fall records). They may be encountered in mixed flocks with yellow-rumped and orange-crowned warblers in May and again in late August–September.

Yellow warblers are not as restricted to the Carson River corridor during migration as they are in the breeding season. They prefer willow thickets or cottonwood overstory for nesting, and are more commonly encountered as breeders in the small creek drainages of Nevada's mountain ranges, as well as along the Truckee, Walker, and Humboldt Rivers, and the Carson River above Lahontan Reservoir. Alcorn (1988) reported a significant decrease in their numbers since 1943, probably due to a decline in the quality of the valley's riparian habitat.

Yellow Warbler

Chestnut-sided Warbler
(*Dendroica pensylvanica*)
STATUS: Vagrant.

One sight record of a male from June 7, 2000, in Fallon (WH).

Magnolia Warbler
(*Dendroica magnolia*)
STATUS: Vagrant.

A single sight record, from Timber Lakes on September 5, 1998 (RF, LN, MN), is the only report of this species in the valley.

Black-throated Blue Warbler
(*Dendroica caerulescens*)
STATUS: Vagrant.

There is 1 sight record, October 7, 1986, from Stillwater NWR (AB 41:123; *American Birds* mistakenly reported 2 birds instead of 1). There are additional sight records from Fernley (Lyon County) and Wadsworth (Washoe County).

Yellow-rumped Warbler
(*Dendroica coronata*)
STATUS: Common migrant; uncommon winter resident.

This species occurs in the valley at almost all times of the year except June and July, when most move to higher elevations to nest. Yellow-rumped warblers are often the most common warblers in migration. Spring migration occurs from the second half of April to early May (May 18, 1995, latest spring record). Numbers in fall migration peak in late August and September. Yellow-rumped warblers will linger in small numbers into November, until the weather gets really cold. Recorded 12 out of 13 years on Fallon CBCs between

1985 and 1997, usually 1–15 birds; exceptions are 34 birds in 1989, 62 in 1994, and 227 in 1997.

Black-throated Gray Warbler
(*Dendroica nigrescens*)
STATUS: Uncommon migrant.

Black-throated gray warblers are mostly likely to be seen along the Carson River corridor during May (May 8, 1998, earliest record; May 17, 1998, latest record). There is an unusual record from June 21, 1997 (RF). Fall migration for this species in the valley is poorly understood. Although there is only 1 record (from September 24, 1998), it is expected that migrants move through in small numbers in August and September. Black-throated gray warblers breed in the pinyon-juniper zone in the Stillwater Mountains.

Townsend's Warbler
(*Dendroica townsendi*)
STATUS: Uncommon migrant.

Townsend's warblers transit the valley in mid to late May and again in August (July 30, 1997, earliest fall record) through early October (October 28, 2000, latest fall record). They are usually seen in the better-developed cottonwood habitats along the Carson River.

Hermit Warbler
(*Dendroica occidentalis*)
STATUS: Vagrant, but possibly overlooked.

There are only 3 records, including a bird collected between Fallon and Hazen on August 17, 1934 (MVZ CA no. 66485). The others are sight records from September 5, 1998, at Timber Lakes (RF); and September 6, 1998, at the Carson River Diversion Dam (RS). This species is a rare breeder in western Nevada's Carson Range. Migrants are regularly seen in the Carson Range in August.

Palm Warbler
(*Dendroica palmarum*)
STATUS: Vagrant.

The 2 valley records include a sight record from November 27, 1987, in Fallon (PL); and a bird banded at Timber Lakes on September 24, 1998 (DW, GA).

Blackpoll Warbler
(*Dendroica striata*)
STATUS: Vagrant.

An adult female was banded and photographed at Timber Lakes on September 8, 1999 (RN, DW).

Black-and-white Warbler
(*Mniotilta varia*)
STATUS: Vagrant.

One was banded at Timber Lakes on August 25, 1999 (RN, DW). This species is probably more common in the valley than the single record indicates.

American Redstart
(*Setophaga ruticilla*)
STATUS: Vagrant.

The 4 records include 2 birds observed on August 31, 1988, at Timber Lakes; 1 seen on January 2, 1997, in Fallon (RF); and 1 banded on September 23, 1998, at Timber Lakes (DW, GA).

Ovenbird

(*Seiurus aurocapillus*)

STATUS: Vagrant.

The single valley specimen was collected June 12, 1941, 4 miles west of Fallon (MVZ CA no. 83232; Alcorn 1941b).

Northern Waterthrush

(*Seiurus noveboracensis*)

STATUS: Vagrant.

There is 1 sight record: September 6, 1998, from the Carson River Diversion Dam (RS).

MacGillivray's Warbler

(*Oporornis tolmiei*)

STATUS: Uncommon migrant.

MacGillivray's warblers pass through the valley in May (May 4, 1959, earliest spring record; June 7, 1971, latest spring record) and again from early August (July 30, 1997, earliest fall record) through the middle of September (September 23, 1994, latest fall record).

Common Yellowthroat

(*Geothlypis trichas*)

STATUS: Uncommon and local summer resident.

Yellowthroats arrive in early May (May 4, 1997, earliest spring record) and remain through mid-September (October 8, 1996, latest fall record). Rather local nesters in tule-cattail stands, they are more common in the freshwater wetlands. They are almost never observed in drains, and seem mostly absent from closed hardstem bulrush stands that lack cattails. Common yellowthroats can be found in portions of Stillwater NWR, Carson Lake, and Harmon Reservoir, and in the wetlands above the Carson River Diversion Dam. In 1898,

Oberholser and Bailey reported this species as common in tule marshes and as occurring in thickets along the Carson River.

Hooded Warbler
(*Wilsonia citrina*)
STATUS: Vagrant.

One sight record: May 10, 1995, from the Carson River Diversion Dam (RF).

Wilson's Warbler
(*Wilsonia pusilla*)
STATUS: Common migrant.

Wilson's warblers are commonly encountered during migration. Spring migration peaks during the first 2 weeks of May (June 1, 1995, latest spring record); fall migration peaks in August with a few birds lingering into September (September 23, 1986, latest fall record).

Yellow-breasted Chat
(*Icteria virens*)
STATUS: Formerly a common summer resident, now a rare migrant.

Migrants are likely to be encountered in early May and in August and early September (September 24, 1998, latest fall record). Alcorn (1988:327) reported chats as "summer residents in the Lahontan Valley seen frequently from May to August.... Since that time [1943], this bird has almost vanished from this area." The yellow-breasted chat is very sensitive to habitat change and is particularly intolerant of the degradation and removal of the willow-buffaloberry mid-story in riparian corridors. The transformation of the riparian habitat in the Lahontan Valley, including the replacement of the riparian mid-story on the lower Carson by Russian olive, seems to account for the absence of breeding yellow-breasted chats. This species remains a breeder on the nearby upper Carson and Truckee Rivers.

FAMILY THRAUPIDAE: TANAGERS

Western Tanager
(*Piranga ludoviciana*)
STATUS: Uncommon migrant.

Migrants pass through the valley during May (May 10, 1998, earliest spring record; May 30, 1996, latest spring record) and again from August (July 30, 1997, earliest fall record) through mid-September (September 20, 1995, latest fall record). Western tanagers are irregularly encountered in the valley in summer. During migration they occur throughout the valley, but particularly along the Carson River corridor where they are attracted to ripe fruit of Russian olives.

Western Tanager

FAMILY EMBERIZIDAE: TOWHEES, SPARROWS, JUNCOS, AND LONGSPURS

Green-tailed Towhee
(*Pipilo chlorurus*)
STATUS: Rare migrant.

Spring records are from the first 2 weeks of May (May 27, 1947, latest spring record), and fall records are from the middle of August through early September (November 3, 1997, latest fall record). Although the green-tailed towhee is common in Nevada's interior mountain ranges, including the nearby Stillwater Range, it is not often observed in the valley floors during migration.

Spotted Towhee
(*Pipilo maculatus*)
STATUS: Uncommon year-round resident.

Spotted towhees are not numerous anywhere in the valley at any time of the year. They are found along the Carson River corridor in riparian habitat, and a few are present in all but the coldest winters. Recorded 9 of 13 years on Fallon CBCs between 1985 and 1997 (1–9 birds per year).

American Tree Sparrow
(*Spizella arborea*)
STATUS: Vagrant; possible rare winter visitor.

Two records, including a specimen from November 25, 1939, about 4 miles west of Fallon (MVZ CA no. 77276); and a sight record from November 27, 1987, at Carson Lake (PL). This species winters regularly in the Honey Lake basin in nearby California and in Elko County, Nevada.

Chipping Sparrow
(*Spizella passerina*)
STATUS: Uncommon migrant.

An early spring migrant through the Lahontan Valley from April (March 21, 1986, earliest spring record) to early May (May 13, 1960, latest spring date), this species breeds in the pinyon-juniper zones in the surrounding mountain ranges, including the Stillwater Range. Fall passage occurs from late August (July 30, 1941, earliest fall record) through September (September 24, 1994, and 1998, latest fall records).

Brewer's Sparrow
(*Spizella breweri*)
STATUS: Common migrant.

Brewer's sparrow is a bird of the true sagebrush zones and is widely distributed as a breeder across Nevada. This species occurs in the valley during migration from the middle of April (April 12, 1997, earliest spring record) to the middle of May (May 24, no year, latest spring record), and again from the middle of August (July 4, 1996, earliest fall record) through September (October 6, 1994, latest fall record). Fall migration peaks from about August 20 through September 10.

Vesper Sparrow
(*Pooecetes gramineus*)
STATUS: Uncommon migrant.

Vesper sparrows occur in the valley from April (March 28, 1988, earliest spring record) into early May and again from late August to September (December 15, 2000, latest fall record). A good place to look for them is along the weedy dikes of Stillwater NWR and Carson Lake and Timber Lakes.

Lark Sparrow

(*Chondestes grammacus*)

STATUS: Uncommon migrant;
rare winter resident and breeder.

Migrants pass through from mid-April (March 19, 1998, earliest spring record) to early May (May 13, 1996, latest spring date) and again from early August (July 30, 1997, earliest fall record) through the middle of September. Lark sparrows have been recorded on 2 CBCS (December 16, 1990, and December 20, 1987), but they may not overwinter. They regularly breed in the valleys of nearby Pershing and Humboldt Counties, and were also confirmed breeding at Stillwater NWR in 1998. An older summer record from June 24, 1965, indicates that this species may have previously bred in the valley as well.

Black-throated Sparrow

(*Amphispiza bilineata*)

STATUS: Uncommon summer resident.

Black-throated sparrows arrive in the valley at the beginning of May (April 20, 1942, earliest spring record) and depart by late September (September 23, 1994, latest record). They are birds of the salt desert scrub, thriving in seemingly inhospitable saltbush-shadscale-greasewood communities. The *Atriplex* dune communities north of Fallon comprise the primary habitat for this species, which does not occur in urban/suburban settings.

Black-throated Sparrow

Sage Sparrow

Sage Sparrow

(*Amphispiza belli*)

STATUS: Common summer resident; uncommon winter resident.

More numerous than black-throated sparrows in basically the same habitats, sage sparrows occur more often around the fringes of developed land and wetlands than black-throated sparrows. Recorded 12 out of 13 years on the Fallon CBCS between 1985 and 1997 (4–41 birds per year).

Savannah Sparrow

(*Passerculus sandwichensis*)

STATUS: Abundant summer resident;
fairly common winter resident.

Along with the song sparrow, this species is the most common sparrow in the valley. Savannah sparrows are birds of open country. Typical habitat includes meadows, cultivated fields, and grassy margins around the wetlands. Recorded 11 of 13 years on Fallon CBCs between 1985 and 1997. It is unlikely that the savannah sparrows breeding in the valley are the same individuals that winter there.

Fox Sparrow

(*Passerella iliaca*)

STATUS: Uncommon migrant;
rare winter resident.

Migrants pass through from the middle of September (August 2, 1997, earliest fall record) through late April (April 23, 1941, latest spring record). A few winter in the valley from November through February. Recorded 4 out of 13 years on Fallon CBCs between 1985 and 1997.

The fox sparrows found in the valley are of the slate-colored, or *schistacea*, subspecies with slate gray heads. This subspecies may receive full species status in the near future.

Song Sparrow

(*Melospiza melodia*)

STATUS: Abundant summer and winter resident;
common migrant.

Song sparrows occur in a variety of habitats from the brushy understory of the Carson River corridor to the wetland's weedy margins.

Song Sparrow

They are commonly recorded on the Fallon CBCs (up to 311 individuals). Alcorn (1988:351) reported specimens from three races: *merrilli*, *fisherella*, and *montana*, "indicating migratory birds from Oregon, Idaho and probably other areas."

Lincoln's Sparrow
(*Melospiza lincolnii*)
STATUS: Uncommon migrant and winter resident.

Migrants pass through the valley in April and again in September (September 2, 1996, earliest fall record) and October. Lincoln's sparrows winter in the valley as well and are probably more common than the records indicate. For example, they were recorded on only 5 of 13 years on Fallon CBCs between 1985 and 1997. This species is found in dense vegetation, particularly in moist areas.

White-throated Sparrow
(*Zonotrichia albicollis*)
STATUS: Rare migrant and winter visitor.

Most likely to be found from October (October 11, 1995, earliest fall record) through late March (May 3, 1996, latest spring record), although in some years a white-throated sparrow can be found during the fall and winter months in the valley. Most records cluster in October, representing fall migrants. This species is most often found in white-crowned sparrow flocks on weedy and shrubby roadside edges.

Harris' Sparrow
(*Zonotrichia querula*)
STATUS: Rare winter visitor and migrant.

When present, this occasional visitor arrives at the end of November and stays through at least January and perhaps later. Spring migrants pass through in April (April 20, 1997, latest spring record).

White-crowned Sparrow

White-crowned Sparrow

(*Zonotrichia leucophrys*)

STATUS: Abundant winter resident and migrant.

White-crowned sparrows arrive in the middle of September (September 7, 1998, earliest fall record) and are gone by the beginning of May (May 9, 1942, and 1995, latest spring records). The high CBC count of 4106 birds was made in 1988. Both the white-lored (wintering) and black-lored races visit the valley.

Golden-crowned Sparrow

(*Zonotrichia atricapilla*)

STATUS: Rare but regular winter resident and migrant.

Migrants appear in the second half of September and peak in October. The small number that overwinter in the valley are supplemented by spring migrants in March through early May (May 3, 1996, latest spring record). The majority of golden-crowned sparrows encountered in the Lahontan Valley are immature birds.

Dark-eyed Junco

(*Junco hyemalis*)

STATUS: Common winter visitor.

Juncos arrive in the middle of September (September 12, 1994, earliest record) and remain through the middle of March (March 16, 1945, latest record). They are found throughout the valley in scrubby and riparian habitats, as well as on the edges of gardens. The juncos found in the valley are mostly "Oregon type," with a few "slate-colored."

McCown's Longspur
(*Calcarius mccownii*)
STATUS: Vagrant.

There are 2 sight records: 14 birds, March 28, 1988, at Carson Lake (AB 26:469); and 8–12 birds from November 23–28, 1999, through at least January 22, 2000, at Carson Lake (JD, PL, MOB).

Lapland Longspur
(*Calcarius lapponicus*)
STATUS: Vagrant.

The 2 sight records include a November 20, 1989, observation at Stillwater NWR of a calling bird (WH); and 1–2 birds at Carson Lake from November 23–28, 1999, through at least January 22, 2000 (JD, PL, MOB).

FAMILY CARDINALIDAE:
GROSBEAKS AND BUNTINGS

Rose-breasted Grosbeak
(*Pheucticus ludovicianus*)
STATUS: Vagrant.

There are 2 sight records: May 15, 1991, from Pine Road in Fallon (WH); and May 2, 1997, from Wildes Road in Fallon (WH).

Black-headed Grosbeak
(*Pheucticus melanocephalus*)
STATUS: Common summer resident.

Black-headed grosbeaks arrive during the first week of May and summer along the river corridor as well as in the older, more established rural residential neighborhoods where ornamental trees have

Black-headed Grosbeak

reached maturity. They usually depart fairly early in August, with a few lingering into September (September 23, 1998, latest record).

Blue Grosbeak
(*Guiraca caerulea*)
STATUS: Rare local summer resident.

The blue grosbeak has been expanding its range into northwestern Nevada since 1992. Individuals are encountered with increasing frequency in the Lahontan Valley and along the upper Carson and lower

Truckee Rivers. A population seems fairly well established in Mason Valley as well. Blue grosbeaks arrive after the first week of May (May 6, no year, early spring record) and depart in September (September 20, 1995, latest fall record). They are most common along the Carson River corridor. Breeding is suspected but not confirmed.

Lazuli Bunting
(*Passerina amoena*)
status: Common migrant; uncommon summer resident.

Lazuli buntings arrive in the valley the first week of May and depart by mid-September (September 11, 1997, latest record). They are uncommon breeders in willow thickets around the valley, and are not necessarily restricted to the Carson River corridor. Lazuli buntings sometimes maintain territories in agricultural drains with sufficient willow cover.

Lazuli Bunting

FAMILY ICTERIDAE:
BLACKBIRDS, ORIOLES, AND ALLIES

Bobolink
(*Dolichonyx oryzivorus*)
STATUS: Vagrant.

There are 3 sight records: a male on July 7, 1997, at Stillwater NWR (RF); a male on May 21, 1998, along Stillwater Slough (WH), and a female on September 1, 2000, at the Carson River Diversion Dam (LN).

Red-winged Blackbird
(*Agelaius phoeniceus*)
STATUS: Abundant year-round resident.

Red-winged blackbirds establish territories in mid-March and have completed breeding and begin flocking in mid-August. Numbers on the Fallon CBCs range from 1338 (1994) to 15,291 (1990). They nest in the wetlands as well as smaller marginal wetland habitats throughout the valley.

Western Meadowlark
(*Sturnella neglecta*)
STATUS: Common year-round resident.

Meadowlarks winter in loose flocks rather locally around the valley. They begin singing as early as mid-February, but serious nesting does not start until April. Fallon CBC totals range between 80 (1993) and 783 (1988).

Red-winged Blackbird

Western Meadowlark

Yellow-headed Blackbird
(*Xanthocephalus xanthocephalus*)
STATUS: Abundant summer resident;
rare / irregular winter resident.

Males usually arrive in March. Breeding begins when females arrive
in April. This species generally leaves by September, but will overwinter in fair numbers in some years. Recorded 8 of 13 years on Fallon CBCS between 1985 and 1997. Aside from nesting by the hundreds
in the wetlands, yellow-headed blackbirds nest along agricultural
drains throughout the valley.

Yellow-headed Blackbird

Rusty Blackbird

(*Euphagus carolinus*)

STATUS: Vagrant.

One sight record on April 29, 2000, of a male from Carson Lake (LN, JP).

Brewer's Blackbird

(*Euphagus cyanocephalus*)

STATUS: Common summer resident;
abundant winter resident and migrant.

Nesting occurs singly, usually on or near the ground. Brewer's blackbirds use a variety of ornamental trees and shrubs. Summer birds are less conspicuous than the winter flocks, which can be found throughout agricultural areas and the wetlands. Fallon CBC totals range from 1160 (1989) to 11,543 (1990).

Common Grackle

(*Quiscalus quiscula*)

STATUS: Vagrant.

Alcorn (1940:170) reported a common grackle found dead on April 14, 1938, 4 miles west of Fallon. The bird "was given to Mrs. Anna Bailey Mills who agreed with me as to the identification of the species. She intended to prepare it as a skin but the bird was 'slipping' and therefore discarded. It was found dead along with about nine blackbirds that had apparently been killed with poisoned rolled oats that were distributed in the area for ground squirrels." This record falls within the dates of a number of California records, including one from Mono County. In addition, there is a photograph of a common grackle taken in June 1998 in Fernley. The nesting record reported in *American Birds* from August 19, 1987, should be ignored as it referred to great-tailed grackles (AB 41:1470).

Great-tailed Grackle *

(*Quiscalus mexicanus*)

STATUS: Rare summer resident; very rare winter resident.

This species may be increasing, though it has not spread as quickly as was predicted. The first record occurred at the Canvasback Club in 1984. Since then, a small nesting colony (<10 pairs) occupied Harmon Reservoir in 1993–94, but was not seen much in 1995. Great-tailed grackles usually arrive in March (February 22, 1993, earliest record) and depart by mid-November, although they overwinter in Elko County and we should expect a similar occurrence in the Lahontan Valley if numbers build. Thirty-three were seen on January 6, 1998, at Carson Lake (WH).

Brown-headed Cowbird

(*Molothrus ater*)

STATUS: Common summer resident;
uncommon to rare winter resident.

This species typically arrives in the valley around the third week of April and leaves by mid-October. Recorded 8 of 13 years on Fallon CBCS from 1985 to 1997; usually 1–5 birds; exceptions are 200 birds in 1991 and 152 in 1993. The large CBC totals are suspect and may include misidentified Brewer's blackbirds.

Hooded Oriole

(*Icterus cucullatus*)

STATUS: Vagrant.

A male was observed coming to a bird feeder in Fallon from May 3 through June 14, 2000 (LN).

Bullock's Oriole

(*Icterus bullockii*)

STATUS: Common summer resident.

This species typically arrives the first week in May (March 26, 1992, earliest spring record) and departs by early September (September 11, 1997, latest fall record). Bullock's orioles are summer residents of the river corridor and mature suburban neighborhoods. They prefer closed-canopy stands for nesting, but will sometimes build nests in solitary trees.

FAMILY FRINGILLIDAE: CARDUELINE FINCHES AND ALLIES

Gray-crowned Rosy Finch

(*Leucosticte tephrocotis*)

STATUS: Vagrant.

The single sight record from the valley was on September 11, 1997, at Timber Lakes (RF, LN). This species may be of regular occurrence in the mountains surrounding the Lahontan Valley, but it is rare at low-elevation sites in western Nevada.

Cassin's Finch

(*Carpodacus cassinii*)

STATUS: Irregular fall and spring migrant; possible winter visitor?

The Cassin's finches found in the Lahontan Valley are passage birds, occurring irregularly from September through April (May 6, 1997, latest spring record). Most records are from the second half of March through April and September. There is a summer record from July 1, 1941, from Fallon (NHM no. 29059).

House Finch

House Finch

(*Carpodacus mexicanus*)

STATUS: Abundant year-round resident.

House finches occur near houses and in riparian areas. From 1985 to 1992, numbers on the Fallon CBCs ranged from 1 to 37 birds. More birds have been recorded recently (e.g., 168 in 1993, and 132 in 1994). This increase might be related to a greater number of observers on recent Christmas counts.

Red Crossbill

(*Loxia curvirostra*)

STATUS: Rare visitor, possibly not detected.

There are 4 sight records: May 4, 1997, in Fallon (RF); July 18, 1919, (Mills) location not noted; July 13, 1996, along the Carson River downstream of Lahontan Dam (RF, LN); and September 15, 1996, in Fallon (BH).

Pine Siskin

(*Carduelis pinus*)

STATUS: Irregular visitor.

Pine siskins are likely to show up in the Lahontan Valley (especially at feeders) just about any time of the year, although the records cluster from the third week of May through late June. Additional records include January 16, 1982, and October 7, 1986.

Lesser Goldfinch

(*Carduelis psaltria*)

STATUS: Uncommon winter resident and migrant.

Migrants pass through from late March through May (June 18, 1997, latest spring record) and again in late August through October. Recorded 8 of 13 years on Fallon CBCs between 1985 and 1997 (25–164 birds). Lesser goldfinches breed in riparian areas just outside the valley along the Carson and Truckee Rivers and may breed in the Lahontan Valley.

Lesser Goldfinch

American Goldfinch
(*Carduelis tristis*)

STATUS: Common migrant and winter resident.

Winter birds remain until mid-May (May 17, 1942, latest spring record) and reappear in late September (September 23, 1986, earliest fall record). Napier reported them as "abundant in migration in Fallon in times to coincide with seed pods on the elm trees, but before leaves erupt" (Alcorn 1988:387). Recorded on 12 of 13 Fallon CBCs between 1985 and 1997. The low count of 9 in 1989 followed the high count of 364 in 1988. American goldfinches are often observed at bird feeders.

Evening Grosbeak
(*Coccothraustes vespertinus*)

STATUS: Irregular visitor.

Most records are scattered from November (November 26, 1944, earliest record) through June, but there are records from July and October; not found every year. This species occurs in small flocks and is usually heard flying overhead. Like cedar waxwings, evening grosbeaks feed on Chinese elm buds in April.

FAMILY PASSERIDAE:
OLD WORLD SPARROWS

House Sparrow
(*Passer domesticus*)

STATUS: Nonnative; abundant year-round resident.

There is limited information as to when this species invaded the Lahontan Valley. The first specific record is a specimen collected by Alcorn on September 30, 1939 (MVZ CA no. 81667). In 1946 Alcorn reported them as abundant. Linsdale (1936) reported that A. K. Fisher had sighted this species in Lovelock in 1908. Fisher also visited Fallon in 1908, but his bird list did not include a house sparrow. If this species was present in Lovelock at that time, however, it may also have been present in Fallon. Ryser (1985) stated that by 1888 this species was probably present over most of the Great Basin, except perhaps for the west-central and southwestern edges. Fallon CBC totals range from 926 (1987) to 6131 (1992); most years, totals fall in the 900–1500 range.

HYPOTHETICAL, ESCAPEES, AND FAILED INTRODUCTIONS

Magnificent Frigatebird
STATUS: Hypothetical.

Anna Bailey Mills reported a magnificent frigatebird flying over Fallon in the 1930s (Alcorn 1988).

Marabou Stork
STATUS: Escapee.

A marabou stork was reported at the Canvasback Club September 9–14, 1974 (Alcorn 1988).

Greater Flamingo
STATUS: Escapee.

One was observed at Stillwater NWR on December 3, 1965. There is an additional sighting from October 21, 1966, also at Stillwater NWR (Alcorn 1988).

White-tailed Kite

STATUS: Hypothetical.

There are several undocumented reports from Fallon, including a white-tailed kite seen at Carson Lake in August 1970s (NS) and one seen May 26, 1988 (ST), at Stillwater NWR. The latter was recorded the same day as the sighting of the Mississippi kite in Fallon. There are a number of January and February records for this species from Truckee Meadows in the early 1970s, as well as a January 26, 1972, record from Yerington (Lyon County) (AB 26:635).

Northern Bobwhite

STATUS: Escapee or failed introduction.

Despite repeated introductions over the years, this species has not become established in the valley (Alcorn 1988).

Scaled Quail

STATUS: Escapee or failed introduction.

Alcorn (1988) reported sighting a single bird near Fallon in September 1933.

South Polar Skua

STATUS: Hypothetical.

Alcorn (1988) reported that Alex Williams found a leg and a band of this species at Soda Lakes on January 1, 1988.

Glaucous-winged Gull

STATUS: Hypothetical.

Alcorn (1988) reported that Wesley Baumann sighted 7 birds at Lahontan Reservoir during the summer of 1979 but gave no further details. The number of late fall and winter records of glaucous-winged gulls and glaucous-winged X western gull hybrids in western Nevada is increasing. There are no confirmed summer records. It is possible that this species will be found in a winter gull flock in the Lahontan Valley.

Black-capped Chickadee
STATUS: Hypothetical.

Evenden (1952) reported seeing several black-capped chickadees in the municipal park in Fallon on October 10, 1950. At the time he was not aware of how unusual the sighting was and presumably did nothing to substantiate it. Since the only other records for this species in Nevada are from Elko County, we are disregarding this record. There are no records of black-capped chickadees from the Sierra Nevada in California or Nevada.

Yellow-billed Magpie
STATUS: Hypothetical.

A single bird was reported on May 13, 1989. There is simply no pattern of vagrancy for this species, and the description is not conclusive (AB 43:516, 44:471).

Summer Tanager
STATUS: Hypothetical.

A female was reported from the Fallon NAS Nature Trail on May 1997 with no further details. This species should be considered a possible vagrant in the valley in light of other records north of the Mojave Desert in Nevada.

3.
Birding Sites in
the Lahontan Valley

The Lahontan Valley is best known for its wetlands, particularly Still-water Marsh and Carson Lake. This chapter highlights these two outstanding areas but also introduces the birder to a number of other fine sites.

What to bring: When birding in the Lahontan Valley, particularly at the more remote sites, remember to bring food and water and to have a full tank of gas. During the summer bring insect repellent, sunscreen, and a hat and expect hot temperatures.

CARSON LAKE

The Carson Lake wetlands offer some of the best birding in north-ern Nevada. This large wetland complex, located at the southern end of the Lahontan Valley, is managed by the Nevada Division of Wildlife. Large areas of relatively uniform shallow water range from 0.5 to 12 inches in depth with expansive stands of wetland veg-etation. There are 5 primary wetland units, as well as an adjacent pasture area grazed by Newlands Project farmers. Currently, wet-land acreage varies with the depth of the winter snowpack in the upper Carson River basin and with the availability of agricultural drain flows. A recent effort by the state of Nevada, The Nature Con-servancy, and the Nevada Waterfowl Association to purchase water rights is helping stabilize water inflows and ensure that wetland habitat is available.

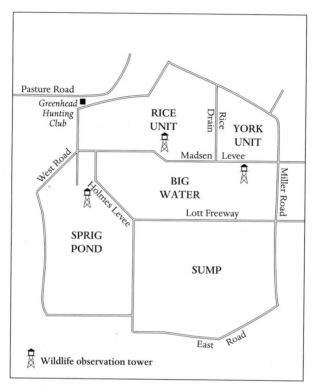

Pasture Road

Greenhead ■
Hunting
Club

RICE
UNIT

Rice
Drain

YORK
UNIT

West Road

Madsen Levee

Miller Road

BIG
WATER

Holmes Levee

Lott Freeway

SPRIG
POND

SUMP

East Road

Wildlife observation tower

Carson Lake

Carson Lake hosts one of the West's largest white-faced ibis colonies. In addition, great egrets, great snowy egrets, and cattle egrets nest along with black-crowned night herons. In years of plentiful water, Franklin's gulls and Forster's and black terns are uncommon nesters. During migration, thousands of waterfowl, including snow geese and smaller numbers of Ross' geese and white-fronted geese, are present along with gadwall, northern pintails, green-winged teal, and cinnamon teal. Peregrine falcons can occasionally be seen hunting shorebirds during migration peaks, which occur in the third week of April and again in the third week of August. Thousands of migrant American avocets, black-necked stilts, long-billed dowitchers, western sandpipers and least sandpipers, long-billed curlews, and many other species create a spectacle unrivaled in the state. Ruffs, ruddy turnstones, red phalaropes, and stilt sandpipers are among the rare shorebirds that have been sighted here. Winter brings large numbers of raptors, including rough-legged hawks, short-eared owls, and bald eagles.

Holmes Levee, Lott Freeway, and Madsen Levee provide excellent viewing opportunities, depending on water conditions. There are 3 viewing towers—1 on Holmes Levee and 2 on Madsen Levee.

Carson Lake is open from sunrise to sunset most days of the year. During waterfowl-hunting season, vehicles are restricted from certain areas. The caretakers can be reached at (775) 423-3071. Visitors to Carson Lake should check in with the caretakers at the gate.

Directions: From the corner of Highway 50 (Williams Avenue) and Highway 95 (Taylor) in downtown Fallon, go 8.8 miles south to Pasture Road. Turn east (left) on Pasture Road and go 2.0 miles to an entrance marked with a "Greenhead Hunting Club" sign. Note: the state of Nevada plans to relocate the main entrance to Highway 95 just south of Pasture Road if the transfer of this area to the state occurs in the near future as planned.

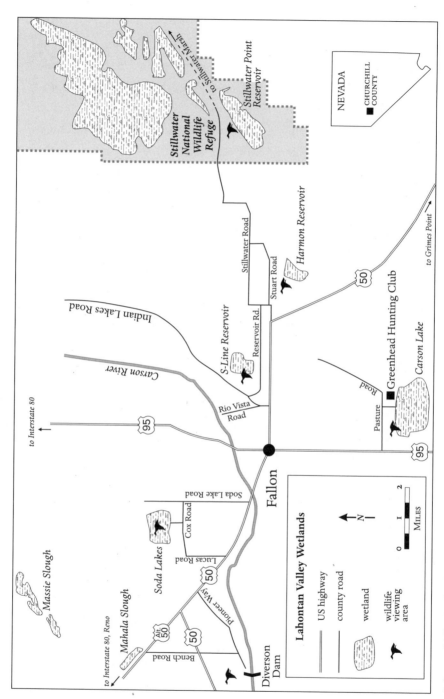

Lahontan Valley Wetlands

STILLWATER NATIONAL
WILDLIFE REFUGE

Along with Carson Lake, Stillwater Marsh is the core of the Lahontan Valley wetlands. Lying to the northeast of Fallon along the foot of the Stillwater Mountains just south of the Carson Sink, one of Stillwater NWR's most precious gifts is solitude.

The marsh is a series of wetland units covering more than 25,000 acres in wet years. The water-rights acquisition program administered by the U.S. Fish and Wildlife Service and The Nature Conservancy is helping to guarantee Stillwater's future. The goal is to acquire sufficient water rights to sustain an average of 14,000 acres of wetlands over the long term.

Stillwater NWR is similar to Carson Lake in many respects, including the large number of waterfowl and shorebirds that visit during migration. Stillwater is particularly noted for attracting large numbers of tundra swans and canvasbacks.

If you are unfamiliar with Stillwater NWR, a visit to the refuge office in Fallon to learn more about water conditions and the wetland locations may save you a lot of driving. Some of the best birding is along the levee separating the Nutgrass unit from Goose Lake. While the first-time visitor may find it difficult to bird in Stillwater, a little persistence yields great results.

Stillwater NWR is open year-round. The Stillwater NWR office can be reached at (775) 423-5128.

Directions from Fallon: From the corner of Highway 50 (Williams Avenue) and Maine Street in Fallon head east on Highway 50. Drive 4.6 miles and turn left onto Stillwater Road. Follow Stillwater Road 12.3 miles to a sign marking the entrance to the refuge.

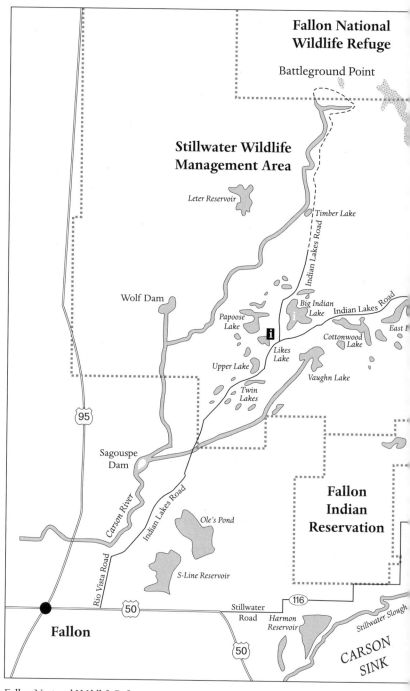

Fallon National Wildlife Refuge

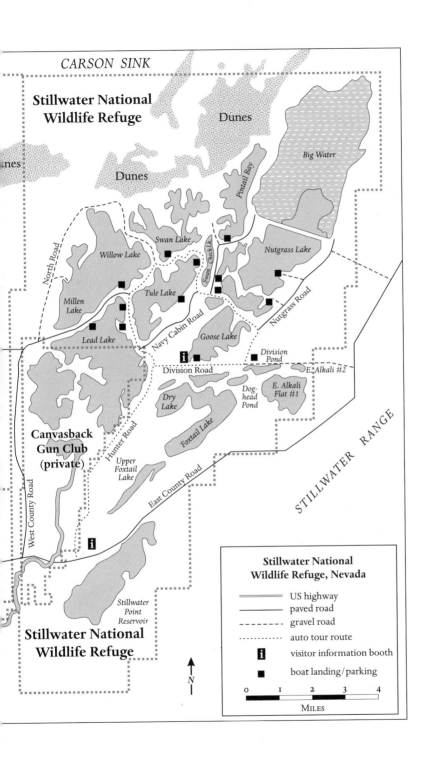

CARSON SINK

**Stillwater National
Wildlife Refuge**

Dunes

Dunes

Dunes

Big Water

Pintail Bay

North Road

Swan Lake

Willow Lake

Swan Check Lk.

Nutgrass Lake

Millen
Lake

Tule Lake

Navy Cabin Road

Nutgrass Road

Lead Lake

Goose Lake

Division
Pond

E. Alkali #2

Division Road

Dog-
head
Pond

E. Alkali
Flat #1

Dry
Lake

Foxtail Lake

STILLWATER RANGE

**Canvasback
Gun Club
(private)**

Hunter Road

Upper
Foxtail
Lake

East County Road

West County Road

Stillwater
Point
Reservoir

**Stillwater National
Wildlife Refuge**

N

**Stillwater National
Wildlife Refuge, Nevada**

	US highway
	paved road
	gravel road
	auto tour route
i	visitor information booth
■	boat landing/parking

0   1   2   3   4

MILES

INDIAN LAKES

The shallow lakes and wetlands northeast of Fallon that comprise Indian Lakes provide interesting water and land birding. The wetlands and lakes currently benefit from a mixture of drain water and fresh water that is diverted to the area. This situation could change in the future and thereby change habitat conditions in this area, although there are several natural wetlands in this area. Indian Lakes attracts a mixture of water birds during migration, including a variety of diving ducks (lesser scaup, ring-necked duck, common goldeneye), as well as mergansers. In addition, during good water years large numbers of grebes, herons, egrets, and terns use Indian Lakes. During fall migration, flocks of land bird migrants can be found in the cottonwood, willow, and tamarisk stands in this area. During the summer months Indian Lakes receives heavy recreational use.

Directions: From the corner of Highway 50 (Williams Avenue) and Maine Street in Fallon head east on Highway 50. After 1.1 miles turn left (north) onto Rio Vista Drive. After 0.4 miles veer right on Indian Lakes Road (the cemetery will be on your right). Follow Indian Lakes Road 6.6 miles to a fork and bear right. Go 1.0 miles to the first lake on the left. From here you will pass a series of lakes and wet areas for the next 3.8 miles.

TIMBER LAKE AND LOWER CARSON RIVER

Beyond Indian Lakes and just upstream of the Carson River's terminus at the Carson Sink is the valley's prime "vagrant trap." Although presently in a degraded condition, the cottonwoods, willows, and Russian olives provide cover for migrants, particularly during the fall. Timber Lake is fed by an artesian well that creates a shallow permanent marsh. During the winter months, up to 50 bald eagles have been reported roosting in the cottonwoods. Please do not disturb the eagles by walking through the grove while they are present. Timber Lake is one place in the valley to see nuthatches, chickadees, and sapsuckers. During summer months, look for western bluebirds and ash-throated flycatchers.

Directions from Indian Lakes: Take northern fork of Indian Lakes Road approximately 6 miles until you reach a dense stand of cottonwoods on your left. Access to the forest is gained through a cattle guard near the north end.

This is a remote area. Please travel with caution and bring food and water.

SODA LAKES

Lying northwest of Fallon and surrounded by arid desert sandhills, Big and Little Soda Lakes were formed by cataclysmic volcanic blowouts. The lakes are a magnet for vagrant shorebirds, gulls, and other birds, and can be the most exciting place to bird in the Lahontan Valley. The list of unusual birds includes Pacific loon, long-tailed duck, surf scoter, white-winged scoter, curlew sandpiper, golden plover, and Sabine's gull. During migration it is not unusual to find large numbers of grebes and phalaropes on Big Soda Lake, attracted by large masses of brine flies and brine shrimp. This is also a good place to study shorebirds, including western sandpipers, least sandpipers, and occasionally semipalmated sandpipers; snowy plovers; dunlins; and dowitchers. Fresh enough to support fish, Little Soda Lake is a good place to look for loons, grebes, diving ducks, and terns.

There are several good vantage points on Little Soda Lake, but Big Soda Lake is most easily viewed from a spit of land on the southwestern shore. The largest diversities of species congregate in a shallow bay on the lake's southwestern corner and on the western part of the lake.

The Soda Lakes took their present-day deep water form after the introduction of large-scale irrigation in the Lahontan Valley at the turn of the century. The remnants of a salt works still exist at the bottom of Big Soda Lake.

Directions from Fernley: From the Highway 50A railroad crossing just east of Hazen, follow Highway 50A 9.1 miles to Lucas Road. Turn left (north) onto Lucas Road. After 1.0 mile Lucas Road becomes Cox Road. Go an additional 1.1 miles and turn left onto a dirt road. You will reach the south rim of Big Soda Lake in 0.3 miles. Turn left and go 0.5 miles to Little Soda Lake from the top of the hill.

Directions from Fallon: From the corner of Highway 50 (Williams Avenue) and Allen Road in Fallon, go west 3.1 miles to Soda Lake Road. Take a right (north) and go 2.0 miles to Cox Road. Turn left onto Cox Road and go 0.9 miles, then take a right onto a dirt road.

CARSON RIVER DIVERSION DAM

One of the valley's best passerine migratory sites is the Carson River Diversion Dam, where birders can gain access to the river and its riparian habitat. Check the area just below the dam for songbirds, particularly warblers and sparrows. White-throated sparrows and Harris' sparrows have been found on the south side of the river channel. In addition, a walk along the south side of the river upstream of the dam will yield a most enjoyable search for passerines in the cottonwoods and Russian olives, and for water birds in the wetlands created by the dam.

Directions: From Highway 50 (Williams Avenue) and Allen Road in Fallon go 7 miles west and turn left (south) onto Pioneer Way. Watch the river and farm fields along the way for birds. After 4.1 miles turn left (west) just prior to a ditch crossing. Follow this gravel road and you will reach the river crossing in 0.3 miles. The river crossing may be impassable during high runoff releases; if so, please respect it and come back another day. If the crossing is passable, cross it and pull over. Check the river crossing for migrants. To continue, bear left and cross the bridge. Note: a gate may prevent vehicle traffic in the near future. Turn right immediately after the bridge. Follow this unimproved dirt road upstream along the river. Park and walk. When you return to Pioneer Way, take a left and in 0.5 miles you will reach Highway 50.

S-LINE RESERVOIR

A regulatory reservoir for the Newlands Irrigation Project, S-Line Reservoir covers approximately 120 acres, though during wet years the northern unit is also flooded, adding up to 320 acres of open-water wetlands. S-Line is usually worth a brief visit, particularly during migration, when grebes, ducks, and small numbers of shorebirds as well as other water birds roost and feed here. A heron and egret colony can be found in the cottonwoods on the northern half of S-Line, and ospreys have nested in this area as well. During winter months, there is often a gull roost at S-Line, and bald eagles are sometimes present as well.

Directions: From the corner of Highway 50 (Williams Avenue) and Maine Street in Fallon head east on Highway 50; after 1.1 miles turn left (north) onto Rio Vista Drive. Go 0.4 miles before bearing right on Indian Lakes Road (the cemetery will be on your right). Follow Indian Lakes Road 0.7 miles and then turn right on Reservoir Road. S-Line Reservoir will be on your left after 0.5 miles.

HARMON RESERVOIR

On the road to Stillwater National Wildlife Refuge, Harmon Reservoir provides a good combination of open water and wetlands totaling approximately 435 acres. The southern half of Harmon Reservoir floods a remnant portion of the slough that historically connected Carson Lake and Stillwater Marsh. Harmon attracts large numbers of grebes (including a red-necked grebe), waterfowl, pelicans, gulls, terns, and a smaller assortment of shorebirds and marsh-dwelling birds, including rails and bitterns (least bittern has been recorded here). The reservoir's Russian olives and tamarisks also attract passerines, including migratory warblers.

Directions: From the corner of Highway 50 (Williams Avenue) and Maine Street in Fallon head east on Highway 50; after 4.6 miles turn left onto Stillwater Road. Follow Stillwater Road 1.1 miles. Where Stillwater Road curves north (left), go straight onto Stuart Road. Check the cottonwoods along Stuart. In 1.1 miles Harmon Reservoir will be on your right. At the east end of the reservoir turn right (south) onto a dirt road that will take you to the back side of reservoir. To continue on to Stillwater, follow Stuart 1.0 mile past Harmon Reservoir and turn left (north) on Ditch House Lane and go 0.5 miles to Stillwater Road.

FALLON NAVAL AIR STATION
NATURE TRAIL

The naval air station (NAS) at Fallon has developed a nature trail featuring riparian vegetation and desert habitat just east of Fallon. The trail is a good site for spotting songbirds during migration. When irrigation is occurring on the surrounding fields, ibis and other water birds can be found there. In winter, flocks of sparrows are regularly found along the irrigation ditches and drains and in the accompanying vegetation.

Directions: From the corner of Highway 50 (Williams Avenue) and Maine Street in Fallon head east on Highway 50. After 2.2 miles turn right (south) onto Crook Road. Follow Crook Road 1.4 miles until you reach Wildes Road. Immediately before Wildes Road on your left is the gated entrance to the nature trail. Enter and park in the parking lot.

BEACH/MACARI ROAD AREA

This area, tucked between Carson Lake and the Fallon Naval Air Station, is a good place to look for winter raptors, including merlins and rough-legged hawks, as well as northern shrikes. The drain ditches attract large numbers of wintering sparrows.

Directions: From the corner of Highway 95 and Pasture Road go east 5.4 miles to Depp Road. Turn right and go 2.0 miles to Schaffer Road. Turn right and go 0.5 miles, then turn right onto Beach Road. Follow Beach Road for 1.0 mile and then take a right onto Macari Road. Follow Macari Road 1.1 miles and turn left onto Cushman Road. Cushman Road will return you to Pasture Road.

LAHONTAN RESERVOIR

This large reservoir serving the Newlands Project is of interest to birders for its nesting colony of California gulls, ring-billed gulls, and the bald eagle nesting pair. During low-water years, however, the exposed mudflats attract shorebirds. The reservoir attracts loons, grebes, and waterfowl, as well as a concentration of wintering bald eagles.

Directions from Fallon: Take Highway 50 west from Fallon to Leetville Junction. Turn left toward Carson City and go 6 miles to the entrance of Lahontan State Recreation Area. Turn left into the entrance and pay the day-use fee.

Appendix 1

TABLE I

Colony-Nesting Wading Birds in the Lahontan Valley, 1986–1999

SPECIES	1986	1987	1988	1989	1990	1991	1992	1993	1994	1995	1996	1997	1998	1999
White-faced Ibis	2475	2973	2400	3075	4139	—	315	475	796	2777	4965	6555	3291	4112
Great Blue Heron	657	540	20	43	38	—	107	41	53	48	180	161	124	187
Great Egret	195	476	90	152	135	—	148	99	123	100	148	119	105	157
Snowy Egret	270	300	180	113	265	—	232	86	212	307	405	245	340	206
Cattle Egret	0	10	255	145	225	—	55	25	17	33	50	120	85	107
Black-crowned Night Heron	375	152	375	151	233	—	51	40	105	199	331	173	81	100
Double-crested Cormorant	155	210	0	3	24	—	40	40	43	41	72	40	51	—

Source: Nevada Division of Wildlife annual surveys.

TABLE 2

Shorebird Counts—April 1989–1999

SPECIES	1989	1990	1991	1992	1993	1994	1995	1996	1997	1998	1999
Black-bellied Plover	323	256	130	175	155	0	2	42	67	23	9
Snowy Plover	47	6	1	0	0	10	24	2	10	4	15
Semipalmated Plover	680	294	43	20	58	29	20	60	50	1	22
Killdeer	108	43	0	5	10	18	19	49	217	216	26
Black-necked Stilt	166	1241	85	267	175	342	224	4532	7279	1988	949
American Avocet	7990	12,431	6687	4123	322	1426	2268	9110	19,637	5496	11,681
Greater Yellowlegs	49	48	1	41	1	0	13	3	72	3	2
Lesser Yellowlegs	2	7	0	2	0	3	12	3	44	11	10
Yellowlegs spp.	0	0	0	0	0	0	0	0	8	0	0
Solitary Sandpiper	0	0	0	0	0	1	0	2	6	0	1
Willet	14	3	9	13	7	4	8	8	17	7	4
Spotted Sandpiper	9	2	2	0	4	10	2	2	4	18	0
Whimbrel	0	0	0	0	0	0	0	4	0	0	15
Long-billed Curlew	32	44	5	23	54	7	33	30	90	65	6
Marbled Godwit	255	298	240	481	32	2	119	23	218	56	155
Western Sandpiper	12,313	7368	0	121	3352	1452	514	623	432	161	1755

Least Sandpiper	23,962	1008	104	164	134	278	460	500	735	456	562
Least/Western Sandpiper	7339	53,949	18,710	955	337	40	100	9311	2379	5883	8265
Least/Western/Dunlin	0	0	70	21,540	600	600	2600	0	0	4095	11,975
Baird's Sandpiper	0	0	0	0	0	0	0	0	0	4	0
Pectoral Sandpiper	2	0	0	0	0	0	0	0	1	0	8
Dunlin	11,136	6385	3578	601	2130	332	58	1061	282	127	1032
Long-billed Dowitcher	39,284	28,418	10,090	13,921	2644	8131	9216	28,795	12,152	20,875	22,624
Short-billed Dowitcher	0	2	0	1	0	0	0	0	—	0	0
Common Snipe	0	4	1	0	0	0	0	2	12	16	11
Wilson's Phalarope	87	296	114	2	11	247	10	199	108	169	2581
Red-necked Phalarope	30	0	10	0	0	2	0	13	121	30	55
Phalarope spp.	10	0	0	0	0	0	0	0	0	0	0
Total	103,838	112,103	39,880	42,455	10,026	12,934	15,702	54,374	43,941	39,704	61,763
Caspian Tern	3	1	14	—	2	2	2	47	114	0	—
Forster's Tern	12	55	9	—	23	16	1	96	348	63	—
Black Tern	0	0	0	—	0	15	0	2	0	435	—
Bonaparte's Gull	0	11	1	—	7	0	0	13	33	0	—
Franklin's Gull	0	0	0	—	0	0	1	1	3	0	—

Source: Based on spring and fall surveys conducted by the U.S. Fish and Wildlife Service and the Nevada Division of Wildlife.

TABLE 3
Shorebird Counts—April 1989–1998

SPECIES	1989	1990	1991	1992	1993	1994	1995	1996	1997	1998
Charadriidae										
Black-bellied Plover	5	1	0	0	0	0	1	4	0	0
Snowy Plover	0	0	7	2	5	13	28	197	76	48
Semipalmated Plover	17	5	2	8	0	2	34	7	31	9
Killdeer	267	53	11	158	93	121	159	263	645	656
Black-necked Stilt	844	1666	394	457	960	2029	5788	6300	8404	7525
American Avocet	6876	23,638	5820	1935	845	2821	14,908	5167	22,761	24,451
Greater Yellowlegs	0	24	0	16	34	9	27	40	130	105
Lesser Yellowlegs	95	13	1	2	13	12	11	31	31	61
Yellowlegs spp.	9	0	113	262	19	10	9	24	128	68
Solitary Sandpiper	3	0	1	0	0	3	1	0	4	0
Willet	1	0	3	3	0	5	17	1	1	3
Spotted Sandpiper	12	17	4	1	8	9	5	7	25	19
Whimbrel	0	0	0	0	0	0	1	0	1	—
Long-billed Curlew	195	20	0	1	97	22	56	54	14	13
Marbled Godwit	81	226	87	112	126	118	44	117	187	171
Western Sandpiper	1633	810	34	20	0	329	573	747	584	1509
Least Sandpiper	1540	125	11	4	37	265	147	344	495	343

Species										
Least/Western Sandpiper	366	11,865	2240	3367	375	1136	552	716	9607	1710
Least/Western/Dunlin	0	0	0	0	0	0	0	4016	0	10,231
Semipalmated Sandpiper	—	—	—	—	—	—	—	—	1	1
Baird's Sandpiper	0	0	0	1	2	3	8	6	17	27
Pectoral Sandpiper	5	0	1	0	1	0	0	1	0	0
Sanderling	0	—	—	—	—	—	—	—	3	—
Ruff	0	—	—	—	—	—	—	—	1	—
Dunlin	2	25	0	0	0	0	0	3	0	3
Long-billed Dowitcher	29,175	28,820	13,204	4540	6000	18,077	13,083	13,322	19,945	22,248
Short-billed Dowitcher	0	0	0	1	0	0	1	2	0	1
Common Snipe	2	0	0	8	6	2	2	27	80	10
Wilson's Phalarope	454	420	7	810	16	4	2849	1311	1417	863
Red-necked Phalarope	1486	871	987	759	737	216	1578	2028	1418	2802
Phalarope spp.	666	3090	0	0	625	884	447	1270	982	5776
Total	53,734	71,689	22,927	12,467	9999	26,090	40,328	36,006	66,987	78,654
Caspian Tern	10	5	6	—	0	0	0	3	0	0
Forster's Tern	19	0	1	3	13	0	293	239	21	93
Black Tern	35	1	22	6	1	35	203	277	539	260
Bonaparte's Gull	0	26	4	—	0	6	17	16	253	0
Franklin's Gull	0	0	—	—	0	0	51	32	37	17

Source: Based on spring and fall surveys conducted by the U.S. Fish and Wildlife Service and the Nevada Division of Wildlife.

Appendix 2

Fallon Christmas Bird Counts, 1985–1999

SPECIES	1985	1986	1987	1988	1989	1990	1991	1992	1993	1994	1995	1996	1997	1998	1999
Pied-billed Grebe	—	—	—	—	—	2	—	—	—	6	12	14	15	18	9
Horned Grebe	—	—	—	—	—	—	—	—	—	—	—	—	1	1	—
Eared Grebe	—	—	1	—	—	1	—	—	—	1	4	1	7	1	3
Western Grebe	—	—	—	—	—	—	—	—	1	2	6	—	4	—	—
Clark's Grebe	—	—	—	—	—	—	—	—	—	—	—	1	1	—	—
American White Pelican	—	—	—	4	—	—	—	—	—	—	1	—	—	—	2
Double-crested Cormorant	—	—	1	—	—	—	—	—	1	—	—	—	1	1	1
American Bittern	—	—	—	—	—	—	—	—	—	—	—	1	7	1	—
Great Blue Heron	34	53	59	91	17	48	28	9	31	23	37	54	100	61	54
Great Egret	—	1	—	1	1	—	—	—	8	10	17	14	15	17	39
Snowy Egret	—	—	—	—	—	—	—	—	—	—	6	—	—	—	4
Cattle Egret	—	—	—	—	—	—	—	—	5	—	2	—	1	—	—
Black-crowned Night Heron	2	4	8	14	1	2	1	—	1	3	9	33	35	40	53
White-faced Ibis	—	—	—	1	—	1	—	—	150	—	99	91	3	23	10
Turkey Vulture	—	—	2?	1?	—	—	—	—	—	—	—	—	—	—	—
Tundra Swan	2	1911	43	2	55	51	17	—	599	24	69	240	1603	2960	476
Gr. White-fronted Goose	—	—	4	—	—	—	1	1	1	—	—	—	—	—	1
Snow Goose	16	600	13	4	—	9	—	—	—	280	3500	590	601	—	190

Fallon Christmas Bird Counts, 1985–1999 (CONTINUED)

SPECIES	1985	1986	1987	1988	1989	1990	1991	1992	1993	1994	1995	1996	1997	1998	1999
Ross' Goose	—	—	—	—	—	1	—	—	—	2	—	—	1	—	—
Canada Goose	299	1020	299	175	602	592	34	678	342	606	179	429	177	37	23
Wood Duck	7	3	—	21	11	21	51	110	238	305	300	52	103	34	1
Green-winged Teal	2	815	11	279	298	3912	1	—	4900	1805	5834	11	219	3457	873
Mallard	2417	529	548	3788	1486	3520	94	669	2914	4103	9917	510	1315	742	2384
Northern Pintail	2	335	—	344	417	433	1	—	660	550	10,074	6	1175	682	79
Cinnamon Teal	3	—	—	—	9	2	—	—	25	—	48	1	4	7	2
Teal sp.	—	—	—	—	—	—	—	2	—	—	—	—	—	—	—
Northern Shoveler	21	440	11	55	34	665	—	—	282	180	1096	115	305	338	1698
Gadwall	—	42	—	4	18	482	—	—	265	580	2702	—	90	38	9
American Wigeon	2	10	—	—	—	17	6	—	105	215	512	—	1	273	7
Canvasback	—	2	5	11	2	—	4	—	4	—	—	5	—	4	4
Redhead	—	—	16	15	27	4	51	7	41	48	202	152	107	9	71
Ring-necked Duck	—	—	15	3	—	1	1	2	8	11	—	3	1	—	10
Lesser Scaup	20	—	20	—	2	4	2	1	3	22	1	2	11	5	2
Common Goldeneye	—	—	8	—	15	17	2	17	5	9	5	13	—	4	42
Barrow's Goldeneye	—	—	—	—	—	—	—	—	—	—	—	—	—	—	2
Bufflehead	—	—	3	3	7	46	—	5	69	41	25	32	90	3	19

Species															
Hooded Merganser	—	—	—	—	—	5	—	—	1	—	4	—	7	—	1
Common Merganser	—	1	—	8	56	6	33	—	—	4	2	9	—	—	10
Red-breasted Merganser	—	—	—	—	—	—	—	—	—	—	—	2	1	—	—
Ruddy Duck	—	20	16	5	5	4	4	—	62	118	116	20	190	40	6
Bald Eagle	—	1	2	3	2	2	2	—	13	2	3	—	1	1	1
Northern Harrier	24	22	18	48	45	70	38	33	51	93	63	72	157	88	42
Sharp-shinned Hawk	1	2	1	4	4	3	4	1	1	8	2	2	4	3	4
Cooper's Hawk	2	4	3	3	3	1	3	3	—	4	1	1	6	3	4
Northern Goshawk	—	—	1?	—	—	—	—	—	—	—	—	—	—	—	—
Red-shouldered Hawk	—	—	1	1	1	—	—	—	—	—	—	1	3	—	1
Red-tailed Hawk	30	44	68	89	92	62	65	58	92	77	48	64	106	83	68
Rough-legged Hawk	24	9	15	61	33	11	23	16	13	6	9	6	27	4	9
Ferruginous Hawk	1	1	—	2	1	—	—	3	3	4	2	2	4	1	—
Buteo sp.	—	4	—	3	4	—	—	—	—	—	—	—	—	—	—
Golden Eagle	4	1	7	3	8	—	5	5	1	4	—	1	11	2	—
American Kestrel	37	36	57	58	54	49	28	27	49	60	28	37	51	35	25
Merlin	—	2	2	2	2	2	3	2	2	2	1	2	3	2	1
Peregrine Falcon	1	—	—	—	—	—	—	—	—	—	—	—	—	—	—
Prairie Falcon	10	12	24	28	19	13	19	13	16	28	5	8	18	10	10
Wild Turkey	—	—	—	—	—	1	—	—	—	—	—	—	—	—	—
Ring-necked Pheasant	9	5	8	3	2	3	2	—	7	3	4	1	2	—	8

Fallon Christmas Bird Counts, 1985–1999 (CONTINUED)

SPECIES	1985	1986	1987	1988	1989	1990	1991	1992	1993	1994	1995	1996	1997	1998	1999
California Quail	81	81	161	162	134	180	45	166	259	285	101	295	384	171	345
Virginia Rail	—	1	—	—	—	—	—	—	1	—	—	1	13	1	—
Sora	—	—	1	—	—	—	—	—	1	—	—	—	—	—	—
Common Moorhen	2	—	—	10	—	2	1	—	2	—	2	6	4	5	—
American Coot	8	479	87	79	167	1866	131	—	1246	227	3196	792	1036	469	432
Sandhill Crane	—	—	—	—	2	2	—	—	3	—	—	—	—	—	—
Black-bellied Plover	—	—	—	—	—	—	—	—	1	—	—	—	—	—	—
Killdeer	20	24	21	32	16	59	4	—	39	18	42	39	22	11	8
Mountain Plover	—	—	—	—	—	3	—	—	—	2	—	—	—	—	4
American Avocet	—	—	—	—	—	—	—	—	—	1	181	—	—	—	—
Greater Yellowlegs	2	13	5	11	49	14	—	—	118	54	56	2	28	20	43
Lesser Yellowlegs	—	2	—	—	—	—	—	—	2	—	1	—	—	—	—
Spotted Sandpiper	—	—	1	—	—	—	—	—	—	—	—	—	—	—	—
Long-billed Curlew	—	—	—	—	—	3	—	—	3	—	—	—	—	—	—
Marbled Godwit	—	—	—	—	—	—	—	—	8	—	—	—	—	—	—
Western Sandpiper	1	—	—	—	—	—	—	—	2	—	—	—	—	—	—
Least Sandpiper	8	15	25	18	—	21	—	—	7	28	138	19	13	25	6
Dunlin	—	—	—	—	—	—	—	—	50	2	22	—	—	—	5

Species															
Long-billed Dowitcher	—	—	5	—	5	5	—	—	17	—	—	—	—	—	2
Common Snipe	9	6	13	14	14	28	3	8	12	19	6	3	5	—	—
Ring-billed Gull	4	—	61	100	136	366	318	345	142	153	245	171	1055	202	162
California Gull	20	27	15	—	—	93	—	—	—	—	1	1	7	3	1
Herring Gull	—	—	—	21	4	4	1	2	1	31	—	1	—	—	—
Rock Dove	143	319	334	444	513	115	241	356	85	263	144	187	233	227	225
Mourning Dove	71	92	409	172	132	157	132	85	283	111	8	209	118	281	468
Barn Owl	—	—	1	—	—	—	1	—	—	1	1	—	1	—	—
Western Screech Owl	—	—	—	—	—	—	—	—	—	—	—	—	1	—	—
Great Horned Owl	—	2	2	1	1	4	7	3	4	5	1	4	—	3	2
Burrowing Owl	1	—	—	—	—	—	—	—	—	—	—	—	—	1	—
Long-eared Owl	—	—	—	—	—	—	—	—	—	—	—	—	—	—	—
Short-eared Owl	—	—	1	—	—	4	—	—	—	—	—	—	—	—	2
Belted Kingfisher	2	4	1	4	—	1	1	2	7	7	8	15	14	10	8
Downy Woodpecker	—	—	—	2	5	3	2	2	2	1	1	3	4	5	3
Hairy Woodpecker	—	—	1	—	—	—	—	—	—	—	—	—	—	—	—
Northern Flicker	33	28	102	75	60	99	57	68	98	169	50	77	159	84	87
Black Phoebe	—	—	—	—	—	—	—	—	—	—	—	—	—	5	—
Say's Phoebe	—	—	2	2	—	1	2	—	2	2	—	7	2	—	1
Vermilion Flycatcher	—	—	—	—	—	—	—	—	—	—	—	1	—	—	—
Northern Shrike	—	—	2	2	3	—	3	1	—	1	—	—	2	1	—

Fallon Christmas Bird Counts, 1985–1999 (CONTINUED)

SPECIES	1985	1986	1987	1988	1989	1990	1991	1992	1993	1994	1995	1996	1997	1998	1999
Loggerhead Shrike	5	6	19	16	10	10	7	8	12	17	7	11	19	11	12
Western Scrub Jay	1	—	—	2	—	—	—	—	—	—	—	4	—	—	—
Black-billed Magpie	427	444	553	499	421	475	481	499	495	632	308	520	787	489	576
American Crow	8	8	409	23	8	139	6	—	3	2	—	7	5	8	113
Common Raven	47	138	229	426	410	447	26	16	85	104	19	209	94	38	58
Horned Lark	2870	2938	4929	7484	5586	5176	1926	7760	4705	11,100	2604	6715	7295	2264	4537
Tree Swallow	—	8	3	—	1	1	—	—	—	—	—	—	—	—	—
Mountain Chickadee	—	3	9	—	4	—	—	—	—	—	—	27	—	—	—
Bushtit	—	—	10	—	4	—	15	—	3	—	—	2	15	14	18
Brown Creeper	—	—	1	—	—	—	—	—	—	—	—	2	—	—	—
Rock Wren	—	1	3	1	2	—	—	—	3	1	5	—	2	—	—
Canyon Wren	—	—	—	—	—	—	—	—	—	—	—	—	2	—	—
Bewick's Wren	1	8	3	1	3	4	2	—	1	2	2	5	9	7	5
House Wren	5	—	—	—	—	—	—	—	—	—	—	—	1	—	1
Marsh Wren	7	12	20	34	9	24	6	5	35	8	14	26	92	12	3
Ruby-crowned Kinglet	—	—	—	—	1	2	2	—	3	13	2	2	10	3	—
Western Bluebird	—	—	—	—	—	—	—	—	—	2	—	—	6	—	2
Mountain Bluebird	—	2	—	10	1	2	—	14	5	51	—	—	6	—	7

Species															
Hermit Thrush	—	2	1	1	—	—	2	1	—	—	1	—	—	—	—
American Robin	190	62	249	79	10	240	45	79	71	50	368	29	41	52	12
Varied Thrush	—	1	—	1	—	—	—	1	—	—	—	—	—	—	—
Northern Mockingbird	3	—	4	2	—	1	—	—	—	—	—	—	1	—	—
Sage Thrasher	1	—	—	—	—	—	—	—	—	—	—	—	—	—	—
European Starling	12,080	15,669	37,956	2236	4909	14,571	9724	9976	10,508	16,833	7263	17,670	26,228	15,046	5737
American Pipit	70	58	524	24	182	38	311	2	60	126	113	100	34	47	3
Cedar Waxwing	—	—	302	—	18	13	—	2	—	—	3	3	—	—	—
Bohemian Waxwing	—	—	—	—	—	2	—	2	—	—	—	—	—	—	—
Orange-crowned Warbler	—	1	2	—	—	—	4	—	1	5	2	—	—	—	—
Yellow-rumped Warbler	9	34	227	7	57	62	15	3	5	—	34	7	7	1	6
Spotted Towhee	—	—	2	2	4	3	—	—	3	2	1	4	9	—	1
Lark Sparrow	—	—	—	—	—	—	—	—	—	2	—	—	2	—	—
Sage Sparrow	—	9	8	18	21	4	20	19	4	6	29	36	41	4	20
Savannah Sparrow	52	61	368	118	77	96	12	62	94	543	112	27	20	20	—
Fox Sparrow	1	—	—	3	—	—	—	—	—	—	2	3	1	—	—
Song Sparrow	127	134	311	129	122	109	100	129	125	159	64	87	104	82	—
Lincoln's Sparrow	—	1	3	1	3	—	1	—	1	1	—	—	—	—	1
Golden-crowned Sparrow	—	1	—	1	—	4	—	—	—	—	2	2	2	—	—
White-throated Sparrow	—	—	—	—	—	—	—	—	—	—	—	—	—	—	1
White-crowned Sparrow	1790	1196	2203	1263	983	1517	3499	1790	1686	2509	2415	4106	3448	2132	693

Fallon Christmas Bird Counts, 1985–1999 (CONTINUED)

SPECIES	1985	1986	1987	1988	1989	1990	1991	1992	1993	1994	1995	1996	1997	1998	1999
Harris' Sparrow	—	—	—	—	—	—	—	—	1	—	—	—	—	—	—
Dark-eyed Junco	74	43	125	39	125	67	11	50	148	77	3	119	72	17	3
McCown's Longspur	—	—	—	—	—	—	—	—	—	—	—	—	—	—	4
Red-winged Blackbird	1646	1862	2296	3582	2341	15,291	2279	5400	3527	1338	1274	3784	3043	3986	4403
Western Meadowlark	314	216	698	783	240	172	111	152	80	359	53	268	517	306	194
Yellow-headed Blackbird	—	1	—	—	—	6	32	1	20	—	2	29	49	3	1
Brewer's Blackbird	5423	1405	2666	2097	1160	11,543	7294	3402	3428	3233	363	2275	4290	3641	2903
Great-tailed Grackle	—	—	—	—	—	—	—	—	—	—	—	1	—	—	—
Brown-headed Cowbird	3	—	2	1	—	1	200	5	152	—	—	25	—	3	2
House Finch	11	1	3	8	37	11	22	17	168	132	20	42	42	43	42
Lesser Goldfinch	25	30	85	—	164	—	—	—	131	29	—	7	60	7	—
American Goldfinch	—	17	276	364	9	79	226	192	196	132	40	32	173	45	17
House Sparrow	1298	926	993	941	1098	1514	3809	6131	1605	1116	299	836	1005	975	1158

Source: National Audubon Society Christmas bird count data.

Bibliography

Alberico, Julie A. R. 1993. "Drought and Predation Cause Avocet and Stilt Breeding Failure in Nevada." *Western Birds* 24:43–51.

Alcorn, J. Ray. 1988. *Birds of Nevada.* Fallon, Nev.: Fairview West Publishing.

———. 1940. "New and Noteworthy Records of Birds for the State of Nevada." *Condor* 42:169–170.

———. 1941a. "New and Additional Records for Nevada." *Condor* 43:118–119.

———. 1941b. "Two New Records for Nevada." *Condor* 43:294.

———. 1946. "The Birds of Lahontan Valley, Nevada." *Condor* 48:129–138.

American Ornithologists' Union. 1998. *Checklist of North American Birds.* 7th ed.

Anglin, Ron. 1994. "A History of the Lahontan Valley Wetlands." Unpublished manuscript.

Bailey, Vernon. 1898. "Nevada: Carson Lake Valley (Wadsworth, Ragtown and Stillwater) Physiography." Unpublished manuscript in the Smithsonian Institution Archives, Record Unit 7176, U.S. Fish and Wildlife Service, 1860–1961, Field Reports, box 69, folder 14.

Behle, William H. 1971. "In Memoriam: Jean Myron Linsdale." *Auk* 88:760–774.

Billings, W. D. 1945. "The Plant Associations of the Carson Desert Region, Western Nevada." *Butler University Botanical Studies* 7:89–132. (Reprinted in 1980 as Northern Nevada Native Plant Society Occasional Paper no. 4.)

Bowman, Timothy D., Larry Neel, and Steven P. Thompson. 1989–90. "A Guide to Birdwatching in Lahontan Valley." *Churchill County in Focus* 3:66–77.

Bundy, Robert M., Jeffery V. Baumgartner, and Barbara Lyn Donohue. 1996. "Ecological Classification of Wetland Plant Associations in the Lahontan

Valley, Nevada." Final report, submitted to Stillwater National Wildlife Refuge, Fallon, Nev.

Charlet, D. A., S. D. Livingston, H. Powell, R. Bamford, T. Wade, M. M. Peacock, C. R. Tracy, and M. Rahn. 1997. "Floral and Faunal Survey of the Stillwater National Wildlife Refuge, Stillwater Wildlife Management Area, and Fallon National Wildlife Refuge. Lahontan Valley and Carson Sink, Churchill County, Nevada." Progress report to the U.S. Department of the Interior, U.S. Fish and Wildlife Service, Stillwater National Wildlife Refuge, Fallon, Nev., August 15.

Chisholm, Graham. 1994. Statement at Hearing before the Subcommittee on Water and Power of the U.S. Senate Committee on Energy and Natural Resources, Reno, Nev., December 11, 1993. Washington, D.C.: U.S. Government Printing Office.

Cronquist, Arthur, et al. 1972. *Intermountain Flora. Vascular Plants of the Intermountain West, U.S.A.* Volume 1. New York: Hafner.

DeQuille, Dan. 1963. *Washoe Rambles.* Los Angeles: Westernlore Press.

Evenden, Fred G. Jr. 1952. "Additional Bird Records for Nevada." *Condor* 54:174.

Fisher, A. K. 1908. "Birds. Fallon, Nevada. August 7–13 and 21–22, 1908." Unpublished manuscript in the Smithsonian Institution Archives, Record Unit 7176, U.S. Fish and Wildlife Service, 1860–1961, Field Reports.

Fowler, C. S. 1992. *In the Shadow of Fox Peak. An Ethnography of the Cattail-Eater Northern Paiute People of Stillwater Marsh.* U.S. Fish and Wildlife Service, Region 1, Cultural Resource Series no. 5, Stillwater National Wildlife Refuge, Fallon, Nev.

Gaines, David. 1988. *Birds of Yosemite and the East Slope.* Lee Vining, Calif.: Artemisia Press.

Giles, LeRoy W., and David B. Marshall. 1954. "A Large Heron and Egret Colony on the Stillwater Wildlife Management Area, Nevada." *Auk* 71:322–325.

Grayson, Donald K. 1993. *The Desert's Past. A Natural Prehistory of the Great Basin.* Washington, D.C.: Smithsonian Institution Press.

Hall, E. R. 1925. "Nevada Report. Birds. Churchill County, Nevada. May 11 to May 18, 1925." Unpublished manuscript in the Smithsonian Institution Archives, Record Unit 7176, U.S. Fish and Wildlife Service, 1860–1961, Field Reports.

Hallock, Robert J., and Linda L. Hallock, eds. 1993. *Detailed Study of Irrigation Drainage in and near Wildlife Management Areas, West-Central Nevada, 1987–90. Part B. Effect on Biota in Stillwater and Fernley Wildlife Management Areas and Other Nearby Wetlands.* Water Resources Investigations Report 92-4024B. Carson City, Nev.: U.S. Geological Survey, Water Resources Division.

Herron, G. B., ed. *Population Surveys, Species Distribution, and Key Habitats of Selected Nongame Species.* Nevada Department of Wildlife Job Performance Report, FAWR Project W-53-R-14, Study 1, Jobs 1-6.

Houghton, John G., Clarence M. Sakamoto, and Richard O. Gifford. 1975. *Nevada's*

Weather and Climate. Special Publication 2. Reno: Nevada Bureau of Mines and Geology.

Kadlec, John A., and Loren M. Smith. 1989. "The Great Basin Marshes." In *Habitat Management for Migrating and Wintering Waterfowl in North America,* ed. Loren M. Smith, Roger L. Pederson, and Richard M. Kaminski. Lubbock: Texas Tech University Press.

Kerley, L. L., G. A. Ekechukwu, and R. J. Hallock. 1993. "Estimated Historical Conditions of the Lower Carson River Wetlands." In *Detailed Study of Irrigation Drainage in and near Wildlife Management Areas, West-Central Nevada, 1987–90. Part B. Effect on Biota in Stillwater and Fernley Wildlife Management Areas and Other Nearby Wetlands.* Water Resources Investigations Report 92-4024B, pp. 7\–20. Carson City, Nev.: U.S. Geological Survey, Water Resources Division.

Jehl, J. J. 1994. "Changes in Saline and Alkaline Lake Avifaunas in Western North America in the Past 150 Years." *Studies in Avian Biology* 15:258–272.

Lico, Michael S. 1992. *Detailed Study of Irrigation Drainage in and near Wildlife Management Areas, West-Central Nevada, 1987–90. Part A. Water Quality, Sediment Composition, and Hydrochemical Processes in Stillwater and Fernley Wildlife Management Areas.* Water Resources Investigations Report 92-4024A. Carson City, Nev.: U.S. Geological Survey, Water Resources Division.

Linsdale, Jean M. 1936. "The Birds of Nevada." *Pacific Coast Avifauna* 23:1–145.

———. 1951. "A List of the Birds of Nevada." *Condor* 53:228–249.

Marshall, David B. 1951. "New Bird Records for Western Nevada." *Condor* 53:157–158.

———. 1952. "Habitat Types of the Stillwater Marsh and Their Value to Nesting Ducks with Reference to Future Management." U.S. Department of the Interior, U.S. Fish and Wildlife Service, Fallon, Nev.

Marshall, David B., and J. R. Alcorn. 1952. "Additional Nevada Bird Records." *Condor* 54:320–321.

Morrison, R. B. 1964. *Lake Lahontan—Geology of Southern Carson Desert, Nevada.* U.S. Geological Survey Professional Paper 401. Washington, D.C.: U.S. Government Printing Office.

Mozingo, Hugh N. 1987. *Shrubs of the Great Basin.* Reno: University of Nevada Press.

Neel, L. A., and W. G. Henry. 1997. "Shorebirds of the Lahontan Valley, Nevada, USA: A Case History of Western Great Basin Shorebirds." *International Wader Studies* 9:15–19.

Oberholser, H. C., and Vernon Bailey. 1898. "Nevada: Stillwater. Birds. May 2 to May 10, 1898." Unpublished manuscript in the Smithsonian Institution Archives, Record Unit 7176, U.S. Fish and Wildlife Service, 1860–1961, Field Reports.

Osugi, Cathy T. 1973. "Monitoring Program of Wildlife Habitat and Associated

Use in the Truckee-Carson Irrigation District, Nevada. Progress Report No. 1." Report of Wildlife Management Study.

———. 1974. "Monitoring Program of Wildlife Habitat and Associated Use in the Truckee-Carson Irrigation District, Nevada. Progress Report No. 2." Report of Wildlife Management Study.

Osugi, Cathy T., and Mark J. Barber. 1976. "Monitoring Program of Wildlife Habitat and Associated Use in the Truckee-Carson Irrigation District, Nevada. Progress Report No. 3." Report of Wildlife Management Study.

Page, Gary W., Lynne E. Stenzel, W. David Shuford, and Charles Bruce. 1989. "Distribution and Abundance of the Snowy Plover on Its Western North American Breeding Ground." *Journal of Field Ornithology* 62:245–255.

Rosenberg, Kenneth V., Robert D. Ohmart, William C. Hunter, and Bertin W. Anderson. 1991. *Birds of the Lower Colorado River Valley.* Tucson: University of Arizona Press.

Rubega, Margaret. 1997. Report on Research on American Avocets and Black-necked Stilts during the 1997 Field Season at Carson Lake, Nevada. Unpublished report.

Ryser, Fred A. 1985. *Birds of the Great Basin.* Reno: University of Nevada Press.

Slipp, J. W. 1942. "Notes on the Stilt Sandpiper in Washington and Nevada." *Murrelet* 22:61–62.

Small, Arnold. 1994. *California Birds: Their Status and Distribution.* Vista, Calif.: Ibis.

Sperry, C. C. 1929. "Report on Carson Sink (Churchill County) Nevada—Its Duck Food Resources and Value as a Federal Migratory Bird Refuge Site." Unpublished manuscript.

State of Nevada. 1950. Report of the Fish and Game Commission for the Period July 1, 1948, to June 30, 1950, Inclusive. Carson City, Nev.

———. 1953. Report of the Fish and Game Commission for the Period July 1, 1950, to June 30, 1952, Inclusive. Carson City, Nev.

Townley, John M. 1998. *Turn This Water into Gold. The Story of the Newlands Project.* 2d ed. Reno: Nevada Historical Society.

Tuttle, Peter L., and Carl E. Thodal. 1998. *Field Screening of Water Quality, Bottom Sediment, and Biota Associated with Irrigation in and near the Indian Lakes Area, Stillwater Wildlife Management Area, Churchill County, West-Central Nevada, 1995.* Water Resources Investigations Report 97-4250. Carson City, Nev.: U.S. Geological Survey, Water Resources Division.

U.S. Fish and Wildlife Service. 1996. *Final Environmental Impact Statement. Water Rights Acquisition for Lahontan Valley Wetlands. Churchill County, Nevada.* U.S. Department of the Interior, U.S. Fish and Wildlife Service, Region 1, Portland, Ore.

Yardas, David. 1994. Statement at Hearing before the Subcommittee on Water and Power of the Senate Committee on Energy and Natural Resources, Reno, Nevada, December 11, 1993. Washington, D.C.: U.S. Government Printing Office.

Index

Page numbers in boldface refer to illustrations